BRIDGE MAINTENANCE INSPECTION AND EVALUATION

CIVIL ENGINEERING

A Series of Textbooks and Reference Books

1. Bridge and Pier Protective Systems and Devices, *Kenneth N. Derucher and Conrad P. Heins, Jr.*
2. Structural Analysis and Design, *Conrad P. Heins, Jr., and Kenneth N. Derucher*
3. Bridge Maintenance Inspection and Evaluation, *Kenneth R. White, John Minor, Kenneth N. Derucher, and Conrad P. Heins, Jr.*

In Preparation

Salinity and Irrigation in Water Resources, *Dan Yaron*

BRIDGE MAINTENANCE INSPECTION AND EVALUATION

KENNETH R. WHITE
Department of Civil Engineering
New Mexico State University
Las Cruces, New Mexico

JOHN MINOR
Department of Civil Engineering
New Mexico State University
Las Cruces, New Mexico

KENNETH N. DERUCHER
Department of Civil Engineering
Stevens Institute of Technology
Castle Point Station
Hoboken, New Jersey

CONRAD P. HEINS, Jr.
Department of Civil Engineering
and Institute for Physical Science
and Technology
University of Maryland
College Park, Maryland

MARCEL DEKKER, INC. New York and Basel

Library of Congress Cataloging in Publication Data
Main entry under title:

Bridge maintenance inspection and evaluation.

(Civil engineering ; v. 3)
Includes bibliographical references and index.
1. Bridges--Maintenance and repair. 2. Bridges --Inspection. I. White, Kenneth R., [Date] II. Series.
TA1.C4523 vol. 3 [TG315] 624s [624'.283]
ISBN 0-8247-1086-X 80-22577

MARCEL DEKKER, INC.
270 Madison Avenue, New York, New York 10016

Current printing (last digit):
10 9 8 7 6 5 4 3 2 1

PRINTED IN THE UNITED STATES OF AMERICA

PREFACE

A handbook is by definition a reference book that can be handled or consulted easily. This one is no exception. The nature of the problem of evaluating bridge conditions is systematically made clear against a thorough study of the background of bridge types, elements, and failures. Although technically complete, the book is not overly theoretical; it is for the practicing engineer, the bridge inspector at work, and the maintenance foreman.

Although, according to Federal Regulations, Title 23, the individual in charge of the organizational unit that has the responsibility of inspecting, reporting, and inventoring must be a registered professional engineer or equivalent, the person in charge of a bridge inspection team is required to have only a minimum of five years experience in bridge inspection assignments and have completed a comprehensive training course. This book is directed toward the inspector and the crew in the field. The majority of field inspectors generally are not trained engineers, but technicians skilled in bridge inspection. The material in this book provides a valuable background that enables the skilled technician to better evaluate real problems and refer them to proper authority.

The handbook contains sections on bridge materials, mechanics of bridge structures, geometrical considerations, and inspection of the superstructure and the substructure—all in a lay person's language. In addition, also in a simple straightforward manner, there are discussions of movable bridges, truss bridges, girder bridges, inspection equipment, damage evaluation, and reporting systems.

Maintenance inspection consideration of bridges is covered in great but simplified detail. Decks, joints, girders, abutments, piers, pier caps, bearings, connections, watercourses, and pavements are a sample of the individual maintenance items discussed with a basic philosophy of putting the information in an easily handled reference.

The authors express their thanks to those who have given encouragement and advice in the preparation of this handbook. The coauthors deserve special thanks for their efforts, which made this manual a reality.

Kenneth White

CONTENTS

Preface iii

Chapter 1

INTRODUCTION 1
I. Historical Aspects 2
II. Inspection Requirements 4
References 6

Chapter 2

BACKGROUND 7
I. Bridge Types 7
II. Bridge Elements 29
III. Bridge Failures 34
References 36

Chapter 3

BRIDGE MATERIALS 38
I. Timber 38
II. Steel 41
III. Concrete 44
References 46

Chapter 4

MECHANICS 47
I. Loads 47
II. External-Internal Forces 49
III. Reactions, Shears, and Moments 51
Reference 60

Chapter 5

REPORTING SYSTEM 61
I. Numbering Sequence 61
II. Sequence of Inspection 62

III. The Inspector's Notebook 62
IV. Standard Forms and Reports 66
V. Rating of Bridges 68

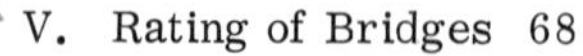

Chapter 6

SUPERSTRUCTURE INSPECTION 71
I. Decks 71
II. Floor Systems 72
III. Beams and Girders 73
IV. Trusses 75
V. Preparation for Inspection 79
VI. Inspection Procedure 82
References 123

Chapter 7

INSPECTION OF THE SUBSTRUCTURE 124
I. Soil-Foundation Interaction 124
II. Abutments 131
III. Dolphins and Fenders 133
IV. Piers and Bents 138
V. Caps 139
VI. Underwater Investigation 141
VII. Culverts 144
VIII. Examples of Substructure Inspection 149
References 155

Chapter 8

MOVABLE BRIDGES 156
I. Introduction 156
II. Types of Movable Bridges 156
III. Operation of Movable Bridges 160
IV. Inspection Considerations 166

Chapter 9

SERVICES 173
I. Signing 173
II. Lighting 175
III. Utilities 178

Chapter 10

INSPECTION EQUIPMENT 181

I. Safety 181
II. Tools and Equipment 185

Chapter 11

BRIDGE CAPACITY RATING 194
I. Deck Capacity Ratings 197
II. Girder Capacity Rating 201
III. Bearing Capacity Rating 220
References 222

Chapter 12

COMPUTER AIDED BRIDGE CAPACITY RATING AND EVALUATION 223
I. Introduction 223
II. Bridge Route Evaluation 223
III. Bridge Capacity Analysis 230
References 232

Chapter 13

WATERWAYS AND TERRAIN 233
I. Introduction 233
II. Problem Areas 233
III. Measuring, Recording, and Detecting Scour 239
IV. Terrain Problems 241

Chapter 14

MAINTENANCE AND REHABILITATION 246
I. Bridge Maintenance 246
II. Bridge Rehabilitation 248
References 251

Index 253

BRIDGE MAINTENANCE INSPECTION AND EVALUATION

Chapter 1

INTRODUCTION

Bridge maintenance inspections during the late 1970s revealed that roadways in the 50 states have just over 1/2 million bridges. Of these 1/2 million, nearly 105,000 of the nation's bridges were rated critically deficient. That is, such structures lack proper safe-load capacity, have poor roadway approach characteristics, have narrow lanes, or have other hazardous features.

Federal Highway Administration (FHWA) engineers have uncovered dangerous bridges in every part of the country. The seriousness of these hazards increases each year, particularly in a year when bitter cold weather causes steelwork to become brittle, resulting in more bridge failures [1].

The FHWA survey shows that

One highway bridge out of every five in the United States is deficient and dangerous to use.

Every two days, on the average, another bridge sags, buckles, or collapses.

Poor bridge approaches and lack of adequate signs or signals annually kill an estimated 1000 Americans, in addition to the 8 or 10 who die as a result of failures of the bridge proper.

Surveying bridge conditions, which may always have been a routine, but small, part of the country's federal-aid highway program, is now a highly organized procedure, and rightfully so, since adequate funding for repair and replacement programs can be developed only on the basis of timely, accurate data substantiating new bridge needs. Programs are now in operation that can provide the data base for federal, state, or local funding programs.

Much of bridge maintenance is often routine, but it also requires a demanding detailed inspection of the highway structures involved

To insure the safety of the traveling public

To protect the initial investment

To detect what is needed for preventive maintenance

I. Historical Aspects

The inspection of highway structures by district highway-maintenance personnel for required maintenance work, if any, has reached a high degree of refinement and organization. All relevant legislation, from the original highway act passed by Congress in 1916 that provided for federal aid to highways to the present, includes requirements for maintenance of the bridges as well as highways.

Since about 1916, inspection of highway structures has been a part, although often a relatively minor part, of maintenance work by the states and has also been conducted by federal employees.

A very detailed inspection program was completed under the Federal Highway Administration (then Public Roads Administration) during the 1930s and 1940s. Maintenance programs following World War II were based in large measure on these detailed reports, thus indicating their value as a tool in fundamental maintenance, quite apart from their contributions to public safety and convenience.

Recent history of bridge maintenance inspections actually began on a Friday, December 17, 1967, when a major bridge collapse occurred on the Ohio River between Point Pleasant, West Virginia and Kanauga, Ohio. Of the 64 passengers and drivers on this structure at the time of its collapse, 46 were killed.

The scope of this tragedy prompted President Johnson to initiate an investigating task force, which met within a week after the collapse. Members formulated three major areas of inquiry:

1. Determining cause of the collapse
2. Determining federal action to accelerate reconstruction
3. Determining procedures available to preclude future disasters and implement changes necessary

Briefly, the first two were accomplished in the following manner:

Individual structure members were reassembled adjacent to the site. Damaged members were examined for possible contribution to the overall failure.

Money was allocated by the federal government for replacement of the structure and design on the new structure was expedited to permit its completion and opening to traffic 2 years from the date of the collapse.

A. Inspections—1968

The committee to evaluate and update safety and inspection procedures was headed by Mister Lowell K. Bridwell, then Administrator of the Federal

Highway Administration, with Mister Alan S. Boyd, Secretary of Transportation, serving as General Chairman. The work of this committee resulted in a March, 1968, memorandum from the Federal Highway Administration directing a review and inventory of all existing highway structures by January, 1970. This instructional memorandum required that pre-1935 structures be reviewed first and that more recent structures be fully attended to later. In addition, all such structures were to be reviewed at least once during each subsequent 5-year period. The more important structures would need inspections every 2 years.

This memorandum also provided for the use of qualified persons in the inspection phase. Actual inspection procedures and techniques were drawn in large part from the 1964 American Association of State Highway and Transportation Officials (AASHTO)* publication, Informational Guide for Maintenance Inspections [2].

With this data and direction at hand, state highway departments began in-depth bridge inspections in mid-1968 following development and printing of appropriate inspection forms.

The immediate result of these 1968-1969 inspections was a complete structure inventory along with identification and correction of the most serious problems. Most important, perhaps, adoption of a detailed, periodic inspection procedure was started on a nationwide basis.

B. Inspections—1970

Shortly after the Federal Highway Administration introduced the 1968 Interim Inspections, Congressional hearings were begun to establish the need for a permanent program of inspections to ensure the public safety. This committee heard testimony from many interested parties, including numerous engineering consultants and various persons from the American Association of State Highway and Transportation Officials, the American Association of Railroads, the American Society of Civil Engineers, and the International Bridge, Tunnel, and Turnpike Association, to name a few. As a direct result of these hearings, the 1968 Highway Act in establishing national standards for bridge inspection, contained specific recommendations for items such as frequency of inspections, and qualifications of inspectors.

Following these hearings, AASHTO and FHWA committees developed the Manual for Maintenance Inspection of Bridges [3], based on congressional requirements, and issued it in 1970. Simultaneously, with preparation of the manual on inspection procedures, the same two agencies developed a training manual for inspection personnel [4].

These two manuals form the basis of the present program promulgated by Congress—one for required procedures, the other for inspector training programs.

---

*At that time, American Association of State Highway Officials.

Because of the importance of the inspection procedures formulated in these manuals, several requirements merit comment in some detail.

II. Inspection Requirements

A. Inspection Personnel

State departments generally administer the federal inspection program through a Bridge Maintenance Engineer, and this engineer is in charge of the teams that perform structure inspections and develop the records and reports in each of the districts. The teams usually vary from three to six members. For each team, a licensed engineer directs the inspection, reporting, and inventory system of structure inspection.

B. Inspection Frequency

A detailed inspection is made every 2 years. Interim inspections may be scheduled at any other frequency to review problem areas. The initial 2-year cycle was to be completed by July, 1973, in all 50 states.

C. Records

The first-time report form included entries for physical inventory purposes—such as a bridge number, dimensions of bridge rail, median roadway width, Defense Highway Section Number, latitude, or design live load. After collection of these facts, survey work and inspection begin. During the inspection, profiles of the deck and channel and a cross section of the channel are obtained. The basic report form in use has divisions for identity and location data, superstructure and substructure condition, planned improvements, survey data, and a final appraisal rating.

The appraisal rating is a numerical "score" for each of the major components of the present structure relative to current desirable characteristics. The components considered here are structural condition, deck geometry, clearances, safe load capacity, waterway, and approach safety.

As a part of the inventory system, inspection data are coded for insertion into a computer. Programs for data storage and retrieval are compatible with the requirements of the Federal Highway Administration to permit their ready access to nationwide statistical data needs and requirements from each of the 50 states.

D. Rating

As a part of bridge evaluation, a permissible operating load is determined for each structure. In most states, charts for the more common types of structure, e.g. pretensioned stringer, simple spans, rigid-slab simple span, have been developed by Bridge Design Sections. These charts are used by the inspector to determine the operating load for a specific structure.

In the past, the Bridge Design Sections have determined whether overload permits should be issued by making a comparison of the moments caused by the overload vehicle to those caused by the standard AASHTO design truck. This method has provided good results; therefore, the charts for use by the field inspector have been prepared on the premise that the operating capacity of a bridge equals the maximum weight a so-called Standard HS Truck can reach before producing the maximum overstress permitted by the Manual for Maintenance Inspection of Bridges [3].

The necessity for posting deficient structures, for weight or speed or both, is developed from the bridge inspector's report. After checking by the Bridge Design Sections and action by the appropriate state agency or commission, appropriate signs are posted.

E. Techniques

1. Survey

Requirements include a channel profile for at least 300 ft above and below the bridge, with elevations obtained at least every 50 ft. Also, a cross section of the channel at the upstream side of the bridge is required, along with a profile of both sides of the bridge deck along the gutter at supports, joints, and midspan. An assumed datum is established from any convenient permanent point, and the work of cross-sectioning and profiling proceeds rapidly. In broad meandering channels, such as the Rio Grande, the channel profile is taken on the meandering thalwet, or actual flowline, regardless of its location.

2. Measurements

The measurements required include roadway and deck widths, deck lengths, span lengths, etc. In addition, numerous items such as clearance (both lateral and vertial), high-water mark location, position of bearings, height of bridge railing, are also required.

One reason for completing measurement and survey phases first is to give the members of the crew an opportunity to become familiar with the bridge and its various parts prior to the inspection phase. Identification of certain types of bearing devices or expansion plates will permit the opportunity to account for problem areas and things to look for, and should, therefore, permit a more concise and accurate inspection.

References

1. "Weak Bridges: Growing Hazard on the Highways," U.S. News and World Report, January 9, 1978.
2. American Association of State Highway and Transportation Officials, Informational Guide for Maintenance Inspections, Washington, D.C., 1964.
3. American Association of State Highway Officials, Manual for Maintenance Inspection of Bridges, Washington, D.C., 1970.
4. U.S. Department of Transportation, Federal Highway Administration, Bridge Inspector's Training Manual 70, Washington, D.C., 1970.

Chapter 2

BACKGROUND

I. Bridge Types

A. Introduction

The inspector must be aware of bridge types in order to properly describe a bridge for an inspection report. Bridge types are classifications not only by type of structure but also by function, as well as by bridge description. All of the above must be properly coded for the structural inventory and appraisal report.

Bridge description implies the function of the bridge with respect to the inventory route and the feature intersected. For instance, a bridge that carries a defense highway and goes over a defense highway gets a special description in the recording and coding guide published by the Federal Highway Administration (FHWA) [1]. Parallel bridges in two directions of travel such as typical interstate bridges also get special note. Bridges described as temporary structures would include Bailey bridges, bridges shored up or those with additional temporary supports, and bridges with temporary repairs to keep the bridge open. These descriptions can be found in the FHWA guide [1].

Classification by function of a bridge pertains to the currently approved classification of the roadway. For roadways, the classifications are: Interstate, other freeway and expressway, other principal arterial, minor arterial, collector, major, minor, and local.

In addition, the custodian or agency responsible for maintaining the structure should be recorded. Examples are: State highway agency, county agency, railroad agency, or federal agency.

Structure data, as contrasted to type of structure, includes year built, lanes on or under the structure, average daily traffic, design load, approach roadway width, median skew (if any), flare, traffic safety features, navigation criteria, whether structure is posted, and type service.

Structure type should be considered carefully with the emphasis on the main span. First the structural material should be noted, as well as a general modifying term, if necessary, to fix an immediate impression with the reader of the report; for instance: Concrete or concrete continuous, steel or steel continuous, prestress concrete or prestress concrete continuous, timber, or masonry.

The nomenclature for design or construction should further describe the type of structure in conjunction with the material. Examples include: Slab, stringer/multibeam or girder, girder and floor beam system, T beam, box beam or girders-single box beam or girders-multiple, frame. orthotropic, truss deck, truss thru, arch deck, arch thru, suspension, stayed girder, moveable-life, moveable-bascule, moveable-swing, or culvert. For the sake of an example, a combination such as steel arch deck puts an immediate image in the reader's mind. Other examples include: Concrete-continuous-box beam single, concrete slab continuous, timber truss thru, or steel continuous girder and floor beam systems. Compare the image of these descriptions with Fig. 2.1-2.5. Additional examples are the steel arch thru, Fig. 2.6; concrete culvert, Fig. 2.7; and prestressed concrete stringer/multibeam, Fig. 2.8.

Approach spans should be labeled as such and carry a similar description as above. For example, the Rio Grande Gorge bridge (Fig. 2.9) has as a main structure description: steel continuous truss, and as deck-and-approach span descriptions: steel, multibeam systems.

The number of spans in the main unit should be recorded, as well as the number of approach spans. The length of each span should be recorded, as well as the total structure length. In a girder, floor beam, stringer system or truss floor beam, stringer system, it becomes important to note the panel span, i.e. stringer span, length since these dimensions can play an important role in quick evaluations of bridge structures for heavy loads.

Fig. 2.1 Steel Arch Deck-Type of Bridge

Fig. 2.2 Concrete Continuous Box Beam Bridge

Fig. 2.3 Continuous Concrete Slab Bridge

Fig. 2.4 Through Timber Truss Bridge

Fig. 2.5 Continuous Steel Girder Bridge

Fig. 2.6 Steel Arch, Thru-Type Bridge

Fig. 2.7 Concrete Box Culvert

Fig. 2.8 Prestressed Concrete Multibeam-Type Bridge

Fig. 2.9 Continuous Steel Deck Truss with Approach Spans

B. Trusses

Bridge trusses can have additional descriptions which date from the "invention" and patent of various types. The three best known are the Howe truss, the Pratt truss, and the Warren truss. These are distinguished by the direction of the diagonals in each panel.

If the diagonals slope downward toward the center (a load in the center produces tension in these members) the truss is called a Pratt truss. Figure 2.10 shows a typical Pratt truss. Careful examination of the truss will reveal that the vertical members are in compression for a downward load at the center.

If the diagonals slope downward away from the center (a load in the center produces compression in these members) the truss is called a Howe truss. Verticals are in compression in the Howe truss.

If the diagonals alternate between downward toward the center and downward away from the center, the truss is known as a Warren truss (see Fig. 2.11). The bridge could have a more complete description as a steel truss, deck of Warren type with two approach spans of steel, and multibeam system.

Fig. 2.10 Typical Pratt Truss Bridge

Larger, more modern trusses could have other configurations. Two of these are the K truss, in which the diagonals form legs of a K, and the Baltimore truss, which has a subtruss configuration within a basic Warren truss (see Fig. 2.12).

Older trusses can have other designations, an adjective such as camel back, indicating a curved upper chord or top member, or Whipple, after a somewhat widely known railroad engineer, or Quadrangular Warren, for a modified Warren truss. Complete descriptions can be found in the Bridge Inspector's Training Manual 70 [2]. Some truss types almost defy common description, such as the concrete truss shown in Fig. 2.13.

As discussed earlier, the roadway position furnishes another descriptive term. A through or thru truss has the roadway going between two trusses with connecting members over the roadway (see Fig. 2.10 and 2.12). If the trusses do not have sufficient height to permit connecting members over the roadway, it is known as a pony truss (see Fig. 2.14). If the roadway is on top of the trusses, the structure is known as a deck truss. The Black River Bridge of Fig. 2.11 is a deck truss.

Fig. 2.11 Typical Warren Truss Bridge

Fig. 2.12 K Truss Bridge

Fig. 2.13 Unusual Concrete Truss Bridge

Fig. 2.14 A Pony Truss Bridge

C. Arches

The arch is one of the older forms of bridge structure. Stone masonry arches were in use between the eighth and fourth centuries B.C. by the Etruscans in what is now north-central Italy. The first iron bridge was an arch built across the Severn near Coalbrookdale in England at a site now called Ironbridge. The key of a true arch is that the structure is in compression only, a state that makes stone or masonry arches possible.

Modern arches can be found in stone or masonry, in concrete, and in steel. Again, one can calssify arches not only by their construction material but also by their shape, roadway position, foundation criteria, and other details.

Typical shapes include the circular, or Roman, arch, the Gothic arch, elliptical arch, and parabolic arch. The actual shape may not be readily apparent to the inspector, but it is worth his while to be precise if the bridge is to have a detailed analysis. Parabolic arches are probably the most common long-span arch, since this shape with a uniform load leads to constant forces within the arch.

Most more modern arches are of concrete or steel. Stone masonry arches are generally of older design and shorter span. The roadway on most concrete and masonry arches is across the top. If the space between the roadway and the arch is solid, it is known as a spandrel-filled arch (see Fig. 2.15). If the roadway is supported by columns to the arch, it is referred to as an open-spandrel arch (see Fig. 2.16).

Steel arches are designed in many configurations. A thru arch has the roadway suspended beneath the arch as shown in Fig. 2.16. Such configurations can have a thru arch with a stiff roadway or deck or a stiff, possibly built-up arch, with a thin, less stiff roadway. In many such cases, the roadway section forms a "tie" between the ends of the arch and the structure is called a tied arch.

A steel arch with the roadway on top normally has columns supporting the roadway. This configuration would be called an open-spandrel arch. An example is the Los Alamos arch of Fig. 2.1. Spandrel-filled steel arches do not exist except in the case of large corrugated steel pipe. Superspans, or long-span structural plates, can be considered as a steel arch although much of the load is carried by arching of the soil above the steel (see Fig. 2.17).

Fig. 2.15 Typical Spandrel Filled Arch Bridge

Fig. 2.16 Typical Open-Spandrel Arch Bridge

Fig. 2.17 Structural Plate-Soil Arch Structure

Fig. 2.18 Through Tied-Arch Footbridge

Timber arch bridges are becoming popular for some situations. Laminated long-span timber arches have been used in both buildings and bridges. An example of a thru tied-arch footbridge is shown in Fig. 2.18.

D. Girders

This section pertains to plate girders, with rolled sections discussed later (although a rolled section technically is a girder). Plate girders can be riveted, welded, and on rare occasions, bolted. They consist of flanges and a web plate. Riveted sections have flange angles to effect a connection between the flange plate and the web plate. Welded plate girders have the flange plate welded directly to the web plate.

Most plate girders have vertical stiffeners that act as bearing posts over load-concentration points and serve as short compression members here, as elsewhere, in the manner of a Pratt truss. The web itself acts as a tension member with the stress or tension field simulating the diagonal in a truss.

Some plate girders, particularly in bridge construction, have horizontal stiffeners located at about one-third of the depth from the compression flange. These horizontal stiffeners prevent the web from buckling due to high compressive flexural stress.

Stiffeners in welded plate girders are plates welded directly to the web plate and possibly to the compression flange. Welds holding stiffeners onto the girder should never cross the weld holding the flange to the web. Such a detail provides a location of reduced ductibility due to a state of triaxial tension caused by residual stresses due to the weld. In short, such details are prime locations for a fracture. If such details exist, they merit close attention during inspection.

Stiffeners on riveted plate girders are of angles riveted directly to the web plate. Bearing stiffeners, i.e., those over locations of concentrated loads, are milled flush with the flange, whereas intermediate stiffeners are not set at such close tolerances.

Girder-type bridges can be simple span or continuous. Depending upon the location of the roadway, the structure can be designated as a through girder or a deck girder. A through-girder bridge is analogous to a pony truss bridge, and a deck-girder bridge, as with a deck truss, has the roadway on the top. Girder bridges with connecting members over the top are rare, although one of the first plate-girder bridges was of this type. The girders of the Brittania Bridge over the Menai Strait designed by Robert Stephenson and built in 1850 are rectangular tubes each carrying one track for railroad trains [3].

Closed tubular plate-girder bridges with the roadway on top are known as box girder bridges. If the top flange, or cover of the box, is made of steel and forms the deck or roadway, the structure is known as a box girder with an orthotropic deck. A recent popular design is a steel-concrete composite box girder that has the top flange and deck made of concrete. Descriptions of these can be found in Ref. 4 and Ref. 5.

Examples of welded and riveted plate-girder bridges are shown in Figs. 2.19 and 2.20.

E. Beams

Simple or continuous beam-type bridges can be made of timber, steel, concrete, prestressed concrete, or other materials. One thinks of a beam-type bridge as having a deck supported by beams that rest directly on supports rather than on floor beams that frame into plate girders or trusses.

Timber beam-type bridges are generally simple span with the deck of timber, or sometimes concrete, supported by the beams or stringers. Such beams can be solid or laminated timber and generally are of relatively short span. Many off-system bridges will be of this type. A typical timber bridge is shown in Fig. 2.21.

Steel beam-type bridges have a rolled section supporting a concrete or timber deck. Such bridges can be either composite or noncomposite. Composite indicates that the design has positive shear connections between

Fig. 2.19 Welded Plate Girder

Fig. 2.20 Riveted Plate Girder

Fig. 2.21 Typical Timber Beam-Type Bridge

the steel beam and a concrete deck. This detail allows the composite section of steel and concrete to act together to carry the load. Older steel beam bridges (prior to 1950s) most likely are noncomposite, that is, the steel beam does not have a positive shear connector between it and the concrete deck. Such a design has the steel beam providing support for all the load by itself, although friction can provide composite action. Such bridges appear stiffer and heavier than those of composite design because of this nonutilized friction. In general, the plans are necessary to determine if a bridge is of composite design. A typical beam-type steel bridge is shown in Fig. 2.22.

Concrete beam-type bridges normally are of two types. One is the T beam bridge which is generally a cast-in-place monolithic deck-and-beam system. The other is the precast, prestressed beam with a cast-in-place deck.

The T beam bridge is named such because of the tee shape used in the analysis of the section. In appearance it has beams supporting a deck, but closer examination will reveal that the deck and beam were cast as one unit, thus enabling the deck to act as additional compressive area for the beam. A typical T beam bridge is shown in Fig. 2.23.

A popular bridge type is the precast, prestressed beam type with a composite deck. These bridges can be found generally as overpasses constructed after the 1950s. In general, an I shape is used for the beam that is an AASHTO standard shape. Prestress indicates that the reinforcing is stressed before loading, thereby placing the entire concrete section in compression.

Such prestressed girder bridges are almost always designed to act with composite action for live load. Construction methods may have used the beams alone to carry the dead load. Some are designed with simple span prestressed beams or girders and continuous decks. Such bridges are considered continuous for live load and simple for dead load.

An example of a typical prestressed beam-type bridge is shown in Fig. 2.24.

Fig. 2.22 Steel Beams

Fig. 2.23 Concrete T Beams

Fig. 2.24 Prestressed Concrete Beams

F. Slabs

A slab bridge is generally continuous, although some simple span slabs exist. Concrete is the preferred material for slab bridges, although glue-laminated timbers are used occasionally.

A slab is nothing more than a wide shallow beam in which the beam itself is used as the deck. Slabs can be of either reinforced concrete or prestressed concrete.

An example of a reinforced concrete, three-span continuous slab is shown in Fig. 2.25.

G. Moveable Bridges

Moveable bridges is a general classification for bridges designed to move to allow traffic on a navigable waterway. Four general types are: (1) the bascule bridge, (2) the swing bridge, (3) the vertical lift bridge, and (4) the floating or barge bridge.

Fig. 2.25 Three-Span Continuous Slab

Bascule bridges pivot about a horizontal axis at one or both ends in order to move out of the way. Such bridges are counterweighted to allow easy movements for opening or closing.

Swing bridges pivot about a vertical axis and swing sideways to clear the navigation path. Such bridges are balanced on the pivot pier to allow movement.

Vertical lift bridges raise as an elevator to allow water traffic to pass underneath. They are easily identified by a tower at each end of the liftable span. Such a bridge is also counterweighted for ease of opening.

A floating bridge is, as the name implies, a barge or floating section of the bridge which is floated out of the way to allow movement of water traffic. Such bridges can pivot about one end or float free.

Moveable bridges require specialized personnel for inspection and maintenance. The FHWA book, Bridge Inspectors Manual for Moveable Bridges [6] and Chapter 8 enumerate the inspection items for these bridges in more detail.

H. Concrete Box Bridges

Concrete box-type bridges are becoming quite numerous and, therefore, require separate discussion. As the name implies, the bridge is a large box or tube section with the roof and floor of the box acting as flanges and the walls (and interior walls) of the box acting as web members. Such structures can be simple span or continuous, single box or multiple box, prestressed or reinforced concrete.

A multiple-box bridge generally is composed of individual precast, prestressed sections set side by side to form the bridge (see Fig. 2.26). These individual boxes are connected together by keys or other means to act as a unit. The top of the boxes, with a simple overlay, forms the deck.

A single box is normally a large cast-in-place unit, of either reinforced concrete, posttensioned or prestressed concrete. The single box may have cells composed of walls in the interior of the box. Outwardly, it appears to be a large single unit. The deck is the top of the box (see Fig. 2.2).

I. Concrete Box Culverts

A culvert is defined as a bridge if the distance from abutment to abutment exceeds 20 ft. Concrete box culverts (CBC) may have individual openings of less than this value but grouped together they form a bridge. A CBC is generally a continuous concrete frame, which is quite popular for use over small or intermittent waterways under fills. An example is shown in Fig. 2.7.

Fig. 2.26 Concrete Box Beams

J. Suspension Bridges

Suspension bridges consist of one or more wire ropes or cables supporting a roadway deck hanging beneath the cables. Very long span bridges generally use this structural system. The Golden Gate Bridge (Fig. 2.27) is probably the best known bridge of this type. In past years this system was popular for shorter span bridges, and many such smaller bridges exist today.

These bridges are easy to identify by the cables draped in a parabolic shape extending over supporting towers at each end of the main span to anchors at the ends of the bridge. The roadway, which is hanging from these cables by vertical cables, can be of truss, plate girder, or box design. A timber-truss, roadway, section suspension bridge is shown in Fig. 2.28.

Fig. 2.27 Suspension Bridge

Fig. 2.28 Timber Truss Suspended from Cables

K. Other

Several other types of bridge structures not discussed above, deserve mention but will not be further elaborated upon here after.

Cable-stayed bridges have a relatively stiff roadway section with additional, but not all, support provided by cables from one or more towers.

Precast concrete sections of various shapes exist as do laminated timber deck bridges.

Bridges of military design include pontoon and Bailey bridges. In most cases these would be considered temporary bridges.

Some states have large, permanent, floating bridges of concrete.

Many strange and wonderful bridges exist—some predesigned, some just made. Many such bridges can be described as combinations of some of the above types. However, no single list of terms can adequately describe every type of bridge.

II. Bridge Elements

A. Superstructure

The superstructure consists of all the parts of the structure that are supported by the bearings on the abutments or piers. The elements of the superstructure include the deck, floor system, supporting members, and bracing. Simple structures such as a concrete slab or laminated timber deck slab carry the load directly to the piers or abutments. The more common setup has the deck supported by longitudinal members, called stringers or girders that carry the load to the supporting system. Larger, more complex bridges such as trusses or plate girders have the stringers supported by a floor beam, which is, in turn, supported by the truss or plate girder.

1. Deck and Floor System

The deck is the part of the superstructure that has direct contact with the vehicle or live load. Decks may be constructed of many materials, the most common being concrete or timber. Steel decks can be found in either open grid decks or orthotropic plate box girders.

Concrete decks normally have the primary reinforcement across the short span direction, which is normally between stringers. Top layers of steel can corrode and cause spalling or delamination of the deck. This problem is normally a result of extensive use of deicing salts coupled with an inadequate covering of the reinforcement.

Concrete slab bridges, or sometimes slabs on trusses without stringers, have the primary reinforcement in the longitudinal direction. If the slab is continuous over a floor beam or pier, large amounts of steel will exist near the top surface over the supporting member.

Timber decks are quite common with older steel truss bridges and with timber girder bridges. The direction of the timber will be across the short span. Treated 2 × 4s or 2 × 6s placed on edge are quite common. Most will have runners or a surface overlay to protect the timber from traffic. These decks can rest on either steel or wood stringers.

Newer laminated timber decks can also be found. Such decks are prefabricated in units 4 or 8 ft wide. Crack patterns in overlays at these intervals are an indication of such decks to the field inspector. Short span bridges may have this type of deck without stringers, which, of course, would have the strong direction between piers.

The complete floor system includes the stringers, which are the longitudinal load-carrying member, and the deck. This system will normally be supported by a floor beam. Fig. 2.29 shows a timber floor system framing into a floor beam. Fig. 2.30 is of a stringerless floor system.

2. Floor Beams

The primary supporting member for a floor system in a truss or plate girder bridge is the floor beam. Stringers normally rest on a frame into the floor beam, which is generally a deep steel section perpendicular to the bridge, spanning the roadway width between individual trusses. Such members have critical connections at the ends, since failure here would normally drop two adjacent panels of the floor system. Floor beams are similar in nature for a truss-type bridge, a plate girder-type bridge, or other types that have the primary supporting unit along the sides of the roadway. A floor beam in a continuous plate girder bridge is shown in Fig. 2.31.

3. Primary Supporting Members

The primary supporting members transmit all loads from the floor beams to the supports at the piers and abutments. Failure of one of these primary supporting members can lead to total collapse of the bridge.

The type of primary supporting member usually identifies the type of bridge. These members can be timber, steel, or concrete beams; steel plate girders; timber, concrete, or steel trusses; arches or steel cables. In general, on large structures the floor system frames into the primary structure through the floor beam at discrete points.

Loads from these points are carried to the supports in similar yet different ways. In a plate girder the bending moment is carried by the flanges —in a simple span, compression on top and tension on bottom. Shear is

Fig. 2.29 Timber Floor System Framing into Floor Beams

carried by the web, which may have vertical stiffeners to help in this function.

Truss type primary supporting members act in a similar fashion but have individual identifiable components. Moment is carried by the upper and lower chord members—top in compression and bottom in tension. Diagonal members carry shear and the verticals serve the same function as vertical stiffness in a plate girder, assuming a Pratt rather than a Howe truss.

Arches have hangers or tension members to the floor beam connections if the arch is above the roadway. If the roadway is on top, columns support the floor system.

Fig. 2.30 Slab Deck Resting on Floor Beams

Fig. 2.31 Floor Beams of a Continuous Plate Girder

Suspension bridges act in a similar fashion with hangers supporting the floor beams from a wire cable in tension. One might look at such a system as an arch in tension supporting a floor system with floor beams and possibly some stiffening members.

4. Bracing

All primary supporting members have secondary members, called bracing, that have functions such as reducing unsupported lengths and holding members in a vertical plane. In trusses these members include portals, cross-frames, and sway bracing. Beam and girder structures have diaphragms or cross bracing, which help stabilize the beams and distribute loads between them.

The term secondary indicates that these members are not primary load-carrying members, and the structure normally will not collapse directly as a result of failure of such a member. It should be noted that although some structures seem to perform adequately with a damaged or missing secondary member, such members serve a purpose and such damage may reduce the capacity or serviceability of the structure as a whole.

5. Miscellaneous

Other components of a superstructure include bearings, expansion devices, railings, sidewalls, parapets, curbs, and deck joints. Elastometric joint material for short span bridges or more elaborate sliding plates or finger joints are considered part of the superstructure. In review, anything above the supporting substructure bearings is considered part of the superstructure.

B. Substructure

The substructure includes those parts of the structure which transfer the loads from the bridge span down to the ground. Two common substructure elements are the abutment and the pier. For a simple single span structure the substructure consists of two abutments. With a longer multispan structure, the substructure would consist of the two abutments and one or more piers.

1. Abutments

The abutment is normally composed of a footing, either spread or pile, a breast wall, a bridge seat, a backwall, and wing walls. The bridge seat is the horizontal portion on which the bridge bearings are situated. The backwall prevents embankment soil from spilling onto the bridge seat. Wing walls, if present, prevent embankment soil around the abutment from spilling into the roadway or waterway that is being spanned.

Sometimes abutments do not have a breast wall and are called spill-thru abutments. As the name implies, the embankment is allowed to spill through the front of the abutment below the bridge seat. These types are readily identified by the earth or berm in front of the abutment.

Those abutments with a large vertical breast wall and wing walls are known as U abutments from the typical U shape in the plan.

The complex nature of the breast wall, wing wall, and footing system makes a detailed analysis complex, so conservative approximations are usually the rule in design. For major structures, a model analysis or finite-element analysis may be necessary. The primary things an inspector checks are movement and crack patterns.

2. Piers

Footings, columns, and caps are the main elements of piers. Footings can be spread, pile, or drilled shaft. Normally, the exact type cannot be determined without the plans. If piles extend to the cap, the common term is bent rather than pier.

Columns transmit vertical load and moment to the footing. If a single column is used, it is sometimes referred to as a stem. A solid wall, instead of individual columns, is common, and the unit is called a solid-wall pier.

The cap is the beam that binds the individual columns into a single unit. Loads from the bridge bearings are distributed to the columns by the cap. The pier cap is most noticeable on rigid-frame piers (two columns) and T piers (single columns or stems).

Material for piers can be concrete, steel, or stone. Wood is common in bents, but steel or concrete piles are not uncommon. River piers generally are more massive than those used on highway underpasses. Starlings, or pointed nosings, are generally used on river piers to reduce the force of water or ice against the piers.

III. Bridge Failures

A. Definition

Failure is defined in the dictionary as (1) a falling short, a deficiency or lack; (2) omission to perform; (3) lack of success; (4) deterioration, decay; (5) bankruptcy; and (6) a person or thing that has failed. An engineer's definition of failure when discussing a structure is the condition existing when the structure fails to perform its intended function. Such is the term's meaning when applied to bridges.

Bridges can fail structurally in a catastrophic fashion, or they can fail through obsolescence. Either results in failure that fits the engineering definition for the state of an object no longer able to perform its intended function.

B. Types of Failure

A recent article in Civil Engineering [7] indicated that 60% of classical bridge failures were caused by natural phenomena including flooding. A total of nine common causes of failure are discussed.

Failures due to flooding account for about half the bridge failures. Generally, one flood can destroy a large number of bridges at the same time, particularly small structures. There are two basic causes of failure related to flooding. One is scour, which can occur without a flood per se; the other is indirect, caused by debris piled against the structure. This debris can reroute the flow, causing scour or create a horizontal pressure that causes failure of the superstructure.

The depth of scour to expect is difficult to estimate. A well-known rule is that four times the difference between flood and low water is equal to the maximum scour to expect. Two references on the problem are Scour at Bridge Waterways, [8] and the FHWA report, Countermeasures for Hydraulic Problems at Bridges [9].

Brittle fracture was listed as the second most frequent cause of failure. The type of failure became frequent as welding first became popular. Better design and control of welding since about 1962 has apparently controlled this problem on newer bridges. Special advice is necessary in nontemperate climates, particularly with nonstandard steel. A classic failure of this type was King's Bridge in Melbourne.

Failures by impact of ships on navigable waterways or trucks at simple overpasses has been increasing in recent years. Examples include the 1971 collision of an 11,000-ton freighter with the U.S. 17 bridge between Jekyll Island and Brunswick, Georgia, which destroyed three, 150-ft spans and the 1978 collision of an "18 wheeler" with the Akela Flats overpass on Route I-10 in New Mexico. Both cases caused prolonged interruption of traffic.

Earthquakes have caused failures in bridges. Recent studies of the 1964 Alaska earthquakes and the 1971 California earthquake have led to proposals for retrofitting existing bridges. The Department of Transportation issued a design manual in 1979 entitled, Seismic Retrofit Measures for Highway Bridges [10]. Most failures due to earthquake fall into one of two types. These types are substructure failure with loss of superstructure-support capacity and superstructure collapse due to excessive relative motion at the support bearings.

Failures have occurred due to wind. The most famous two examples are the Firth of Tay collapse in 1879 and the Tacoma Narrows failure in the early 1940s. The Tay bridge incident was due to excessive wind pressure whereas the Tacoma one was caused by aerodynamic oscillation. Other bridges have been hit by tornados.

Corrosion and fatigue apparently caused the most recent catastrophic failure. Fatigue may be defined as the gradual spreading of a crack under repetitive loading. If the steel has rust or corrosion at the tip of the crack, spreading may be more rapid.

The Point Pleasant Bridge was built in 1929, was successful and in use for many years, and collapsed without warning in 1967. Failure was due to fracture at the pinhole of a single eye-bar. This catastrophic collapse, which cost several lives, in addition to being an example of fatigue failure, points out vividly the need for redundancy in bridge structures.

Although several classic examples are listed above, most common failures of small bridges do not get publicity beyond a local area. Scour, impact damage, and overload are common. The inspector can reduce chances of a bridge failure in his jurisdiction by careful attention to detail in his inspection and analysis. The ability to discern between critical deficiencies and cosmetic deficiencies goes a long way toward preventing bridge failure.

References

1. U.S. Department of Transportation, Federal Highway Administration, Recording and Coding Guide for the Structure Inventory and Appraisal of the Nation's Bridges, Washington, D.C., January 1979.
2. U.S. Department of Transportation, Federal Highway Administration, Bridge Inspector's Training Manual 70, Washington, D.C., 1970.
3. J. P. M. Pannell, Man the Builder, Crescent, New York, 1964, pp. 232-235.
4. U.S. Steel Corporation, Highway Structure Design Handbook, Pittsburgh, Pa.
5. O. W. Blodgett, Design of Welded Structures, The James F. Lincoln Arc Welding Foundation, Cleveland, Ohio, 1966.
6. U.S. Department of Transportation, Federal Highway Administration, Bridge Inspectors Manual for Moveable Bridges, Washington, D.C., 1977.
7. David W. Smith, "Why Do Bridges Fail?," Civil Engineering, American Society of Civil Engineers, New York, November 1977.

8. National Cooperative Highway Research Program Synthesis of Highway Practice, Scour at Bridge Waterways, Topic 5, Highway Research Board, National Research Council, National Academy of Sciences, Washington, D. C., 1970.

9. Federal Highway Administration, Countermeasures for Hydraulic Problems at Bridges, Report No. FHWA-RO-78-162, Washington, D.C., 1978.

10. U.S. Department of Transportation, Federal Highway Administration, Seismic Retrofit Measures for Highway Bridges, FHWA-TS-79-217, Washington, D.C., 1979.

Chapter 3

BRIDGE MATERIALS

I. Timber

Possibly the first bridge crossing in the United States was Coronado's bridge of 1541 which is reported to have been built over the Pecos River in New Mexico [1] . Presumedly, it was a wooden trestle bridge, for timber was the principal bridge material in the early history of the U. S. Wood declined as the principal material of major bridge structures a decade or so before the Civil War. The acme of a large timber structure is considered to be the "colossus" by Lewis Wernwag, built in 1812, which had an arch truss spanning 340 ft over the Schuylkill River near Philadelphia. Many more recent highway bridges of shorter span have been constructed of timber and are still in use. Because of the large number of timber bridges still in use, as well as the newer designs calling for laminated timber, a bridge inspector must become familiar with timber or wood as a structural material.

Wood Structures, A Design Guide and Commentary published by the American Society of Civil Engineers [2] provides an excellent state-of-the-art, including references, for the design of wood structures. Here, our concern is limited to the properties of timber that pertain to inspection and analysis of bridge structures.

A. Durability

Although wood does not corrode, it is subject to attack by fungi, certain insects, and marine boring organisms. Generally timbers used in bridges are treated with chemicals that are toxic to wood-destroying organisms.

Conditions favorable to wood-destroying fungi (decay) include adequate moisture, heat, and oxygen. Localized areas of high moisture content in otherwise dry wood members are susceptible to decay. Examples are untreated end grain of beams and columns. Even if wet conditions are only intermittent, decay can occur; fungi will become dormant during dry periods and active during wet ones.

If the cell cavities of wood are completely occupied by water, as in submerged structures, air will be insufficient for fungi to be active. Wood piles may be too dry in their upper portions and too devoid of air in their lower portions to favor decay. However, the region between the two extremes, such as the surface of the ground or water, presents conditions favorable for decay. Any structure expected to have a degree of permanence should have been treated, although it is often difficult for an inspector to tell what preservative treatment has been used. Decay is likely to occur at connections, splices, support points, bolt holes, or any cuts made in a surface after treatment.

Insects harmful to wood include termites and carpenter ants. Termites include the subterranean termite, the dry wood termite, and the damp wood termite. Generally, damage is inside the wood, and invisible, so the only indication of infestation may be mud shelter tubes or excessive sagging or crushing. Carpenter ants do not feed on wood but build galleries in moist, soft, or partially decayed wood. Debris from the gallery building is generally evident.

Marine borers pose a severe threat to marine piling in sea water. The most likely place for damage is between high and low water marks, although damage may extend to the mud line.

B. Mechanical Properties

Traditionally, many of the mechanical properties of wood have been understood on the basis of tests of small pieces that are considered clear and straight-grained, since they are free of characteristic defects, such as knots, checks, and splits. Allowable levels of stress are derived from such test data, as well as the potential for defects in a particular grade of wood. Determining allowable unit stresses for timber in existing bridges requires sound judgment on the part of the inspector or engineer making the field investigation. The maximum allowable unit stress used in analysis should not exceed 1.33 times the allowable unit design stresses for stress-grade lumber given in the current AASHTO *Standard Specifications for Highway Bridges* [3]. Reduction from the maximum allowable stress depends upon the grade and condition determined at the time of inspection.

The more common mechanical properties are the modulus of rupture in bending, compression strength perpendicular and parallel to the grain, and shear strength. Allowable tension parallel to grain is derived from bending tests. Typical ranges of allowable stresses depending upon species and grade are listed below:

Bending	700—2300 lb/in.^{2}
Tension parallel to grain	325—1500 lb/in.^{2}
Horizontal shear	65— 125 lb/in.^{2}
Compression perpendicular to grain	245— 245 lb/in.^{2}
Compression parallel to grain	425—2150 lb/in.^{2}
Modulus of elasticity	1.0×10^6—1.9×10^6 lb/in.^{2}

In general, wood is strong in bending and compression and tension parallel to the grain, but weak in shear parallel to the grain and compression perpendicular to the grain.

C. Formulae

Typical design and analysis equations for timber members are shown below*:

1. Resisting Moment

$$M = F_b S$$

where, M = bending or resisting moment, F_b = allowable stress in bending, and S = section modulus.

2. Horizontal Shear

$$V = \frac{F_v Ib}{Q}$$

where, V = total vertical shear, F_v = allowable stress in shear, I = moment of inertia, b = width of section, and Q = statical moment of area; or for rectangular section,

$$V = \frac{2}{3} F_v bd$$

where, d = depth of section.

3. Compression (Columns)

$$P = F_c A$$

where, P = total axial load, F_c = allowable stress in compression, and A = area of column, or for long square or rectangular columns,

$$P = \frac{0.30EA}{(\ell/d)^2}$$

---

*These formulae are modified for abrupt changes in section, end conditions, deep members, and other criteria. Consult a design guide for details.

or, for long round columns,

$$P = \frac{3.619EA}{(\ell/r)^2}$$

whichever is less, where, E = modulus of elasticity, ℓ = unsupported length, d = least dimension of column face, and r = least radius of gyration.

II. Steel

Many highway bridges are constructed with primary members of steel. These bridge types include the simple girder, the continuous girder, various types of trusses, and others. In all cases, an understanding of the basic properties of the material is necessary to properly evaluate the structure.

A. Stress-Strain Properties

Many different kinds of steel have been used in bridge construction. All steels can be mechanically evaluated for a few basic properties. These normally include yield stress or strength, tensile strength, and some measure of ductility such as elongation. Allowable stresses are normally based on the yield stress, usually 55% of the yield. Over the years, basic allowable stresses have increased as the expected yield stress increased. Examples of mild carbon steel are listed below.

Year	$lb/in.^2$	Allowable $lb/in.^2$
Prior to 1905	26,000	14,000
1905 — 1936	30,000	16,000
1936 — 1963	33,000	18,000
After 1963	36,000	19,000

After 1963, silicon, nickel, and other high-strength alloy steels were available, in addition to carbon steel. In evaluating steel of unknown specification, one must assume carbon steel with properties associated with the era. Guidelines are available in the AASHTO Manual for Maintenance Inspection of Bridges [4].

One of the advantages of steel is its ability to deform greatly beyond yield without loosing strength. This property allows the use of rivets, holes, and other conditions of high elastic stress without detailed elastic analysis. If steel is overstressed locally, it simply yields and redistributes the stress elsewhere. Note the typical stress-strain curve with attention to the difference between strain to yield versus strain to rupture in Fig. 3.1.

B. Stress Concentrations

Elastic materials such as steel have, due to the physical shape of the material, locations of high local stress. These high stresses occur near notches, holes, or abrupt changes in section. Stresses can be shown both through the theory of elasticity and experimentally, to be as high as three times the normally computed stress. In static cases, this local stress may not be of concern because of the ductility of steel. However, in cyclic loading, these local stress concentrations can pose a problem, as shown below.

C. Fatigue Considerations

Steel, as most materials, will rupture at a stress below yield stress if the stress is applied a sufficient number of times. This phenomena can best be demonstrated by a stress versus number of cycles to rupture, commonly called SN diagram, as shown in Fig. 3.2. As the maximum stress permitted increases, the number of cycles decreases to a value known as the fatigue limit. At the fatigue limit, the material can be cycled through an infinite number of cycles without failure. The problem with stress concentrations should be obvious. Simple stress analysis does not include stress concentrations; therefore, high stresses could be occurring, which could lead to fatigue failure. Newer specifications address this problem directly by having lower allowable stresses for areas of high stress concentration. Older specifications address this problem only indirectly, therefore, inspectors should place a high priority on locating and observing such areas during their routine inspections.

To better develop a feel for the SN diagram shown, the following equivalents are provided:

Cycles	Applications per day for 25 years
20,000 — 100,000	2
100,000 — 500,000	10
500,000 — 2,000,000	50
Over 2,000,000	over 200

In summary, the inspector should know the material with which he is dealing, as well as potential problems and their cause.

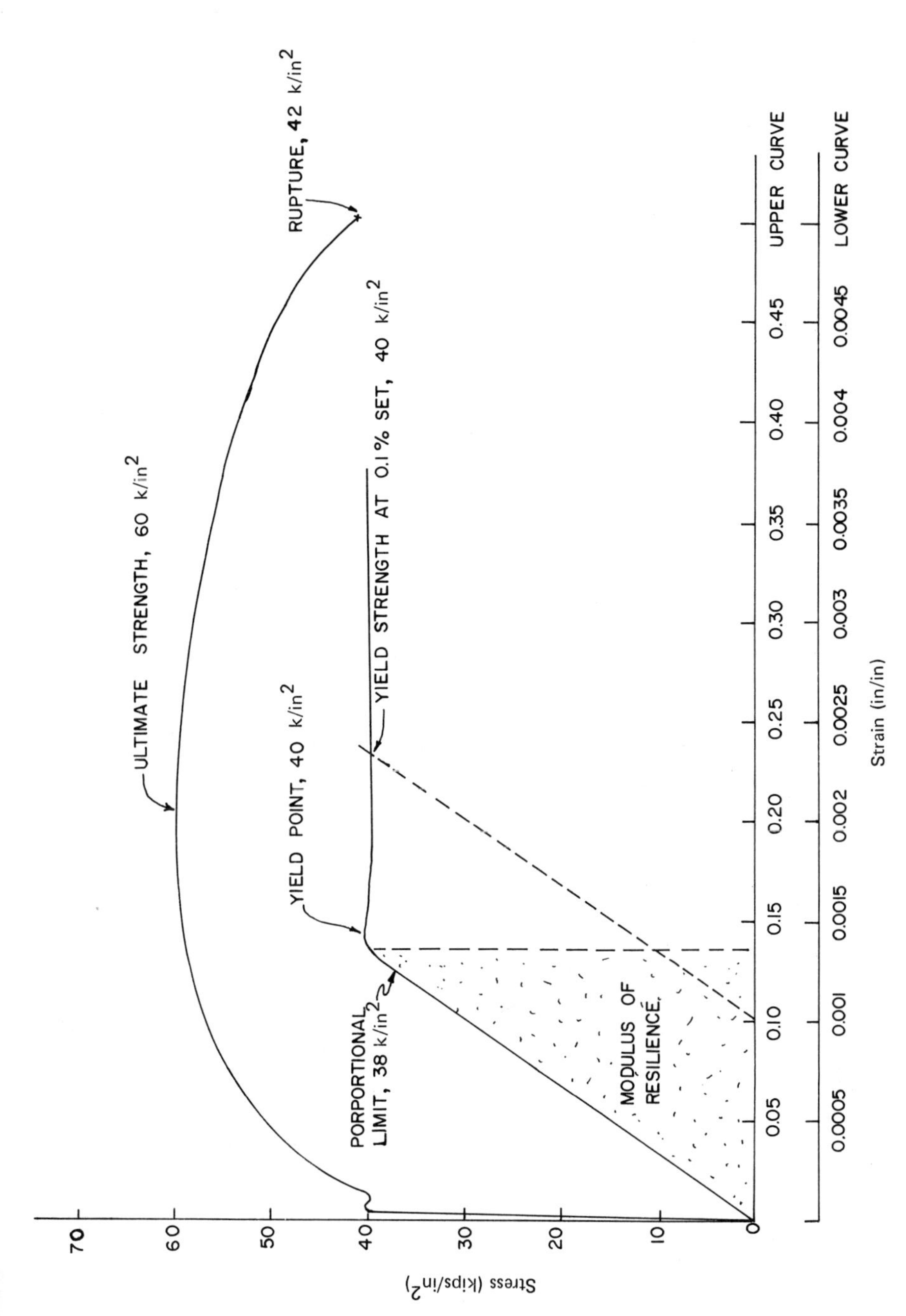

Fig. 3.1 Typical Stress-Strain Curves for Steel in Tension

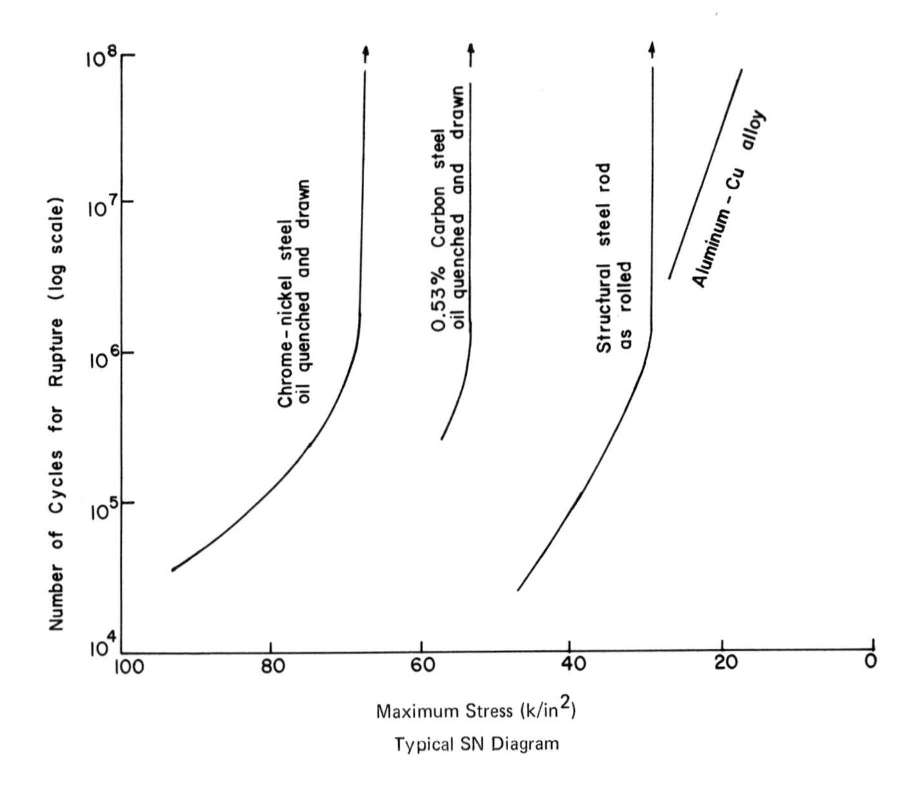

Fig. 3.2 Typical Fatigue Diagram Showing the Relationship of the Stress Level and Number of Load Cycles to Fatigue

III. Concrete

Many highway bridges are constructed with reinforced concrete decks or completely of reinforced concrete. Bridges typically constructed primarily of reinforced concrete include slabs, box beams, T beams, stringers, precast girders, prestressed girders, and box culverts. An understanding of the basic properties of concrete is necessary for the proper evaluation of an existing concrete bridge.

A stress-strain curve for concrete in compression is shown in Fig. 3.3. Critical values include the ultimate strength, the strain at maximum stress, the strain at failure, and the modulus of elasticity. All these values can be obtained from typical stress-strain curves. It should also be noted that concrete is weak in tension, an average stress value being approximately one-tenth the compressive strength.

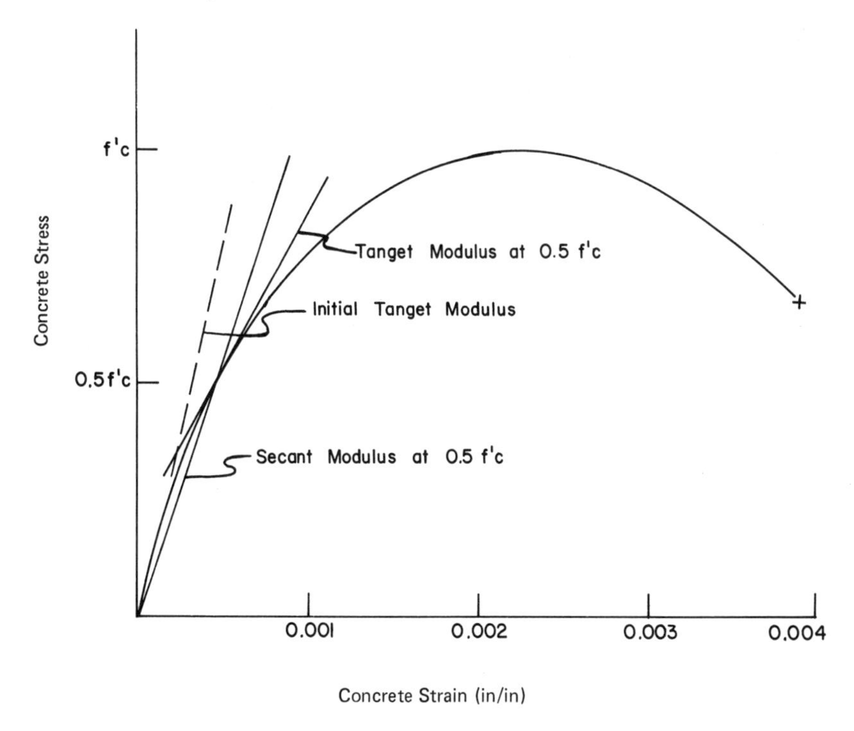

Fig. 3.3 Typical Stress-Strain Curve for Concrete in Compression

Classical analysis shows that reinforced concrete will crack under normal working conditions; however, the use of prestressed concrete should eliminate this problem. Also, the strength of the material is rarely the controlling criteria in the capacity of a bridge, although deterioration can affect strength and reduce the load carrying capacity. Concrete also creeps with age and thereby creates unusual deflections. The deflections will be downward for reinforced concrete, but could be upward for prestressed concrete.

Cracks in reinforced concrete should be noted and an evaluation conducted as to the cause. Typical causes include stress (in design, all potential tension areas are crossed by reinforcement), shrinkage, rusting of steel beneath the concrete, or traffic damage.

The inspector should understand the properties of concrete so that he may properly evaluate apparent problems.

References

1. American Society of Civil Engineers, American Wooden Bridges, ASCE Historical Publication No. 4, New York, 1976.

2. American Society of Civil Engineers, Wood Structures, A Design Guide and Commentary, New York, 1975.

3. American Association of State Highway and Transportation Officials, Standard Specifications for Highway Bridges, 12th Ed., Washington, D.C., 1977.

4. American Association of State Highway and Transportation Officials, Manual for Maintenance Inspection of Bridges 74, Washington, D.C., 1974.

Chapter 4

MECHANICS

The mechanical relation between the applied loads to a member and some quantity of force, usually of stress or deflection, permits a bridge structure to fulfill its load-carrying function. Obtaining a threshold measure of the combined forces permissible on a structure is the bridge inspector's main consideration in understanding the mechanics of bridges. Failure is defined here as any action in a member resulting from applied loads which causes the member to cease to function satisfactorily. A detailed discussion of mechanics is not practical here, and therefore this chapter will be limited to a brief review of loads, force relationships, internal force computations, and deflections.

I. Loads

There are two general types of loadings: dead loads and live loads. Dead loads are the loads that remain on the structure at all times. Total dead load would consist of the combined weight of the various components of the bridge: the deck, girders, bracing, railing, and substructure. In general, anything permanently attached to the bridge is a dead load. Live loads are temporary or changeable loads. Everything that is not a dead load is a live load. The primary live loads on a bridge are the vehicles. Live loads also comprise such things as wind, earthquake, water pressure, thermal forces, ice floes, and similar changing forces on the structure.

Various combinations of these loads may act on the bridge at the same time to produce the maximum loading conditions. An impact factor that accounts for the effect of moving vehicles as they cross the structure is also included in many of the live load computations.

Loads may act on the bridge in various directions. The critical direction is normally vertical and is of primary concern to the bridge inspector. Vertical loads include the weight of the structure, the weight of vehicles, and various lesser combinations such as overlays, pedestrians, water, or snow.

A second manner in which loads act on a bridge occurs in a lateral direction. These loads act from the side and include such things as wind, earthquake, water pressure, ice, and occasionally, collision by vehicles. A lateral pressure such as soil pressure acts selectively on some of the components of the bridge such as abutments, piers, and piling. Normally lateral loads are not the critical load. The bridge inspector should be mindful of such situations and possibilities, however, to ensure computation of the proper load capacity of the bridge. For instance some bridges have tie downs or lateral shear connections for carrying flood-related loads, and such secondary structures should be inspected.

Longitudinal loads normally are considered to be secondary forces and usually include thermal processes, inertia of vehicles, abutment movements, and earthquake. Temperature effects are usually negated by expansion devices at bearings or joints. If these devices cease to function properly, then the longitudinal forces due to temperature change can be quite significant. The inertia effect from the accelerating or braking of vehicles is usually not significant, unless the bridge is on a grade, or sudden alignment changes by roadway alignment changes cause frequent braking in large trucks. Soil pressures on abutments or "growing pavements" may at times exert significant forces on the superstructure. "Growing pavements" occur occasionally when cracks in concrete pavement open up at low temperature and then fill up with dirt and debris. Expansion occurring at high temperatures causes the pavement "to grow" because such expansion is not sufficiently absorbed by the cracks. The expansion, then, is usually transferred to bridge site and exerts tremendous force on the abutments.

The live load of particular concern to the inspector is the standard vehicle, or vehicles, used in determining the load-carrying capacity of the bridge. Most agencies are using the HS20-design truck to maintain uniformity from state to state in rating the capacity of bridges. The H15 and H20 designs are used in some secondary or off-system bridge capacity rating. The "legal truck" configuration is also used, based on the argument that such vehicles give capacity ratings that can be more easily compared to vehicles for which permits are requested.

The H truck has two axles, with 20% of the total load acting on the front axle and 80% on the rear axle. The H15 truck has a gross load of 15 tons and the H20 carries a gross of 20 tons. The HS20 truck has 3 axles incorporated into a tractor truck with a trailer. The tractor weighs 20 tons with 20% of this load on the front axle (4 tons) and 80% on the rear axle (16 tons). The trailer axle carries the same load as the rear tractor axles (16 tons). The gross weight of the HS20 vehicle, then is 36 tons. A more detailed explanation concerning axle spacing, width of wheel lines, and other details may be found in AASHTO Bridge Specifications [1].

The capacity rating of a bridge should include the effect of "impact loadings." The AASHTO specifications use the equation

$$I = \frac{50}{L + 125} \leq 30\%$$

where L is the span length in feet and I is the percentage increase in stress due to impact. The effect of impact may not exceed 30% according to this equation. Impact effect is applied to the superstructure of concrete or metal bridges including rigidly connected columns or portions of the structure extending down to the main foundation. Impact is not applied to timber structures, abutments, piers, retaining walls, foundations, sidewalk loads, or structures having at least 3 ft of soil cover.

II. External-Internal Forces

A force is the action of one body on another body that changes or tends to change the state of motion of the body acted on. External forces are the forces that act upon the bridge elements. The various loads discussed in the previous section make up most of these external forces. Internal forces are the forces which develop in the bridge element to resist the external forces or loads. These internal forces usually fall into one of four categories; axial forces, bending forces, shear forces, or torsional forces.

A. Axial Forces

An axial force acts along the length of a bridge member, producing either compression or tension in that member. A truss bridge utilizes a combination of axial forces in tension members and in compression members to carry the external forces or loads.

An axial force in tension tends to elongate the member. Tension members of a truss are normally long slender members. These tension members normally carry large forces per unit area and therefore the cross sectional examination is of primary importance. Tension members that have lost a part of their area due to damage, deterioration, or fasteners should be inspected thoroughly at such locations.

An axial force in compression tends to shorten the member. Compression members of a truss usually have a much larger cross-sectional area. Compression members are susceptible to buckling, that is, the member snaps out of its plane and loses its load-carrying ability very suddenly. Members that are damaged, deteriorated, or crooked are very susceptible to buckling. The force-per-unit area, or stress, in a compression member is usually much lower than in a tension member because of this buckling problem. Though short timbers or concrete members may crush under large external forces, longer, more slender members will buckle at stresses well below the crushing stress of the material.

B. Shear Force

Transverse forces exert a shearing force, or a tendency to slide a part of a member to one side of a cross section transversely with respect to the other part of the cross section. The larger transverse shears are produced near reactions where large forces may act in opposite directions. Horizontal shear may be produced in a member subject to bending, since one part of the member is in compression while the remainder is in tension. Most design criteria are based on average vertical shear stress. The shear force-per-unit area (shear stress) is determined by dividing total shear force by the web area of a member.

C. Bending Force

A rotational force, or moment, exerts a bending effect on a member in the plane of the member. The units of a moment are the product of a force and a distance. When an external moment is applied to a beam, or member, an internal resisting moment is developed. This internal moment is characterized by longitudinal compressive and tensile fiber stresses acting in the member. The upper portion of the member may be in compression and the lower portion in tension, or vice versa. The stresses are greatest at the upper and lower beam surfaces, and they decline to zero at the neutral axis.

While bending occurs in many structures, it is most common in beam and girder spans. The beam most in use is a simple span. A simple span beam is supported only at its ends and may be a timber, concrete, or steel beam. The dead and live loads act predominantly downward, and the supports resist such action. The result is a bending of the beam, with compression in the top fibers and tension in the lower fibers. A moment producing this type of bending is usually considered positive in bridges. The maximum bending moment on a simple span occurs near the middle of the span.

A continuous beam extending over intermediate supports also produces positive bending; yet, over the supports the action is reversed with the upper fibers in tension and the lower fibers in compression. This is usually termed negative moments and is present in continuous structures and rigid frame bridges.

D. Torsional Forces

Torsion is a twisting of a member about its longitudinal axis. In a bridge, torsion forces are produced primarily in beams by loads on the deck that are not directly over the beam. For instance, as the deck deflects the greatest amount near wheel loads, the beam is twisted by the deck. This twisting produces torsion in the beams. Wind also can cause torsion. Wind forces that act upon one side tend to produce an "overturning force" that produces torsion in some members of the bridge.

The twisting or torsion of a member produces the greatest unit force (torsional stress) at the external surfaces of a member. The stress is zero along the longitudinal axis and increases away from this axis. A closed member, boxed section, has the greatest resistance to torsion and is often used in bridges where torsion may be significant.

III. Reactions, Shears, and Moments

The determination of the forces in the various members of a bridge requires calculations based on the equilibrium of a free-body diagram. The free-body diagram is a sketch of a portion of the bridge showing all external forces acting on the free-body and the resulting internal forces. The external forces are known values and the internal forces are computed using the equilibrium or statics equations

$$\Sigma F_V = 0$$

$$\Sigma F_A = 0$$

$$\Sigma M = 0$$

where F_V = vertical forces

F_H = horizontal forces

M = moments.

The calculations required for continuous-type structures is beyond the scope of this chapter and therefore the discussion is limited to simple-span bridges.

A. Reactions

The reactions, or forces at the supports, must be calculated before the internal forces can be determined. This calculation is accomplished using the equilibrium equations as shown in the examples below. These reaction calculation examples utilize Fig. 4.1.

1. Example 1: Reactions

Determine the magnitude of the reactions at A and B in Fig. 4.1 due to each loading condition indicated. If the concentrated load $P_1 = 100$ lb and acts at the middle of the beam, the reactions are computed as follows. Assume clockwise rotation is positive, then

$$\Sigma M_A = 0$$

$$6(100) - 12\,R_B = 0$$

$$R_B = \frac{6}{12}(100) = 50\text{ lb}$$

$$\Sigma F_V = 0$$

$$R_A + R_B - 100\text{ lb} = 0$$

$$R_A = 50\text{ lb}$$

2. Example 2: Reactions

If $P_2 = 200$ lb and acts alone 9 ft from the left support, then the reactions are

$$\Sigma M_A = 0$$

$$9(200) - 12(R_B) = 0$$

$$R_B = 150\text{ lb}$$

$$\Sigma F_V = 0$$

$$R_A + R_B - 200 = 0$$

3. Example 3: Reactions

If the beam weighs 50 lb/ft length in Fig 4.1 and $P_1 = P_2 = 0$, the reactions are computed as follows:

$$\Sigma M_A = 0$$

$$50(12)6 - 12\,R_B = 0$$

$$R_B = 300\text{ lb}$$

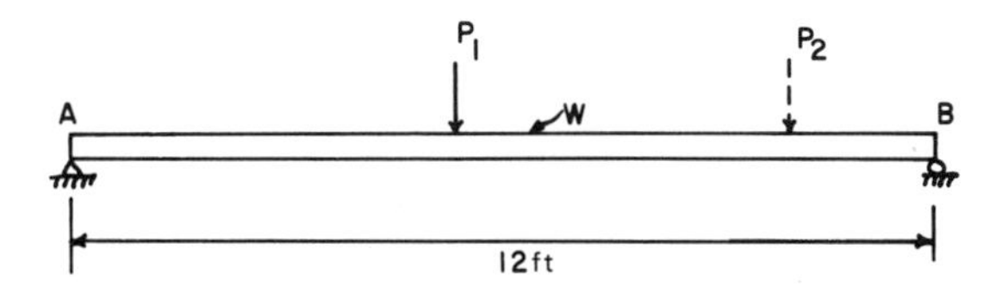

Fig. 4.1 Beam Model Showing Loads for Example 1 and Example 2

$$\Sigma F_V = 0$$

$$50(12) - R_A - R_B = 0$$

$$R_A = 300 \text{ lb}$$

If all the loads act on the beam at the same time, the total reactions may be determined by summing the values computed for each individual load. In this case positive is assumed upward.

$$R_A = 50 + 50 + 300 = 400 \text{ lb}$$

$$R_B = 50 + 150 + 300 = 500 \text{ lb}$$

B. Shear and Moment Diagrams

The quickest method for determining the shear and bending moments in a beam is plotting these values. Once the reactions have been computed, this is a fairly simple operation. A convenient sign convention must first be established, however. The easiest method for shear is to begin at the left support and determine the shear from the applied external loads. If the sum of the external loads at any point on the beam is upward, the shear is positive, a downward sum of loads is negative. Positive bending moments are normally assumed to produce compression in the top fibers of the beam. If the sum of the external loads at any point would bend the beam in a concave upward shape, the moment is positive. The magnitude of the moment can be computed by determining area under the shear diagram. The change in the bending moment is equal to the area under the shear diagram for some reference location (reaction or applied load) to the point in question.

1. Example 4: Shear and Moment Diagrams

Draw the shear and moment diagrams for the loading conditions described in Example 1, a concentrated load at the center of the beam.

The results are shown in Fig. 4.2. The shear diagram was drawn as follows. The left reaction, R_A, was equal to 50 lb upward, so the shear is equal to a positive 50 at point A. The load does not change between point A and the middle of the span or the location of the applied load of 100 lb. A horizontal line represents no change in the shear from point A to midspan. The 100 lb is downward and the net result of the external forces at the midspan is a downward force of 50 lb or a negative shear of 50 lb. The load does not change from this point to point B and therefore another horizontal line is drawn. The reaction at B is upward at 50 lb. The net result of all applied forces is zero at point B.

The bending moment diagram is constructed as follows. The moment at external supports or reactions of simple beams is always zero. Therefore, the moment is zero at point A. Since the change in moment is equal to the area under the shear diagram, the moment at a point 3 ft from point A is

$$\Delta M = 50(3) = 150 \text{ ft}\cdot\text{lb}$$

$$M_3 = M_A + \Delta M = 0 + 150 = 150 \text{ ft}\cdot\text{lb}$$

At midspan, the moment is computed by

$$\Delta M = 50(3) = 150$$

$$M_{\mathrm{CL}} = M_3 + \Delta M = 150 + 150 = 300 \text{ ft}\cdot\text{lb}$$

At point B, the moment is

$$\Delta M = (-50)\ (6) = -300 \text{ ft}\cdot\text{lb}$$

$$M_B = M_{\mathrm{CL}} + \Delta M = 300 - 300 = 0$$

Equilibrium requires that the moment at B be zero since it is an external support for a simple beam.

2. Example 5: Shear and Moment Diagrams

Draw the shear and moment diagrams for the beam shown in Fig. 4.3.

The reactions are determined as follows:

$$\Sigma M_A = 0$$

$$100(3) + 100(7) - R_D (10) = 0$$

$$R_D = 100 \text{ lb}$$

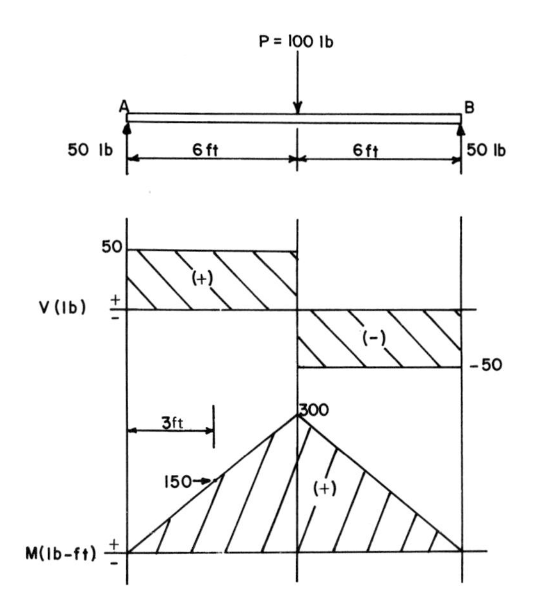

Fig. 4.2 Beam Model and Loads for Example 4

$$\Sigma F_V = 0$$

$$R_A + R_D - 200 = 0$$

$$R_A = 100 \text{ lb}$$

The shear is again determined from the external loads. The reaction at A is 100 lb upward, the load does not change between A and B, hence, a horizontal line or no change in shear is shown. At point B, the applied load of 100 lb produces a net shear of zero between B and C. The applied load of 100 lb at C then produces a shear of -100 lb from C to D and the reaction at D brings the shear back to zero.

The bending moment at A is zero. The change in moment from A to B is

$$\Delta M = 100(3) = 300 \text{ ft} \cdot \text{lb}$$

$$M_B = M_A + \Delta M = 0 + 300 = 300 \text{ ft} \cdot \text{lb}$$

The shear equals zero from B to C and

$$\Delta M = 0(4) = 0$$

$$M_C = M_B + \Delta M = 300 + 0 = 300 \text{ ft} \cdot \text{lb}$$

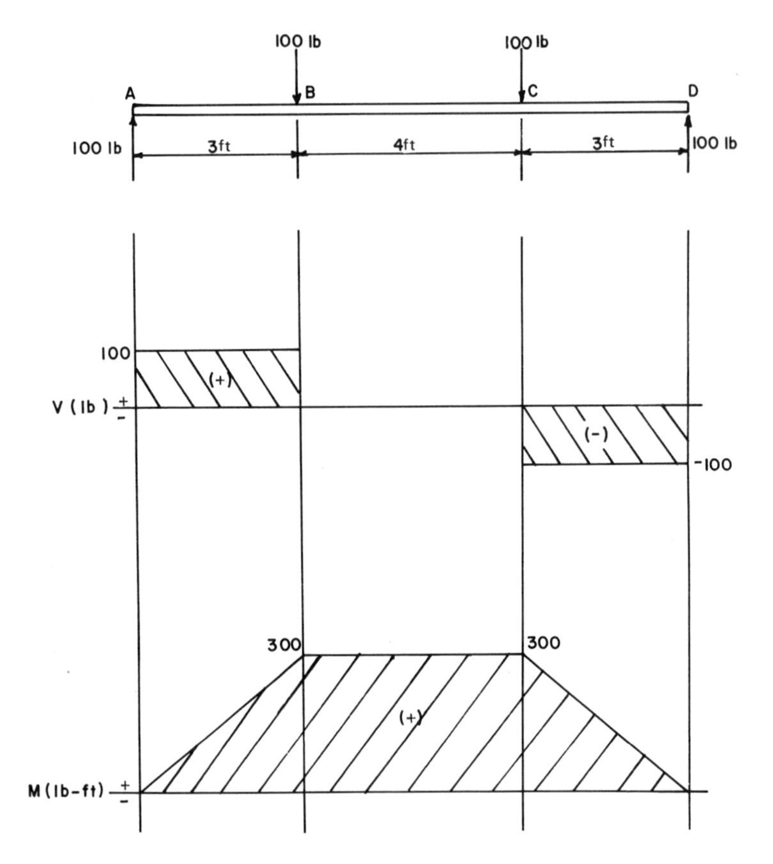

Fig. 4.3 Beam Model and Loads for Example 5

The moment at D is determined as follows:

$$\Delta M = (-100)\ (3) = -300 \text{ ft} \cdot \text{lb}$$

$$M_D = M_c + \Delta M = 300 - 300 = 0 \text{ ft} \cdot \text{lb}$$

Again, equilibrium is satisfied at D with a moment of zero for an external support.

3. Example 6: Shear and Moment Diagrams

Draw the shear and moment diagrams for the distributed load shown in Fig. 4.4.

The computations for reactions are similar to those in the previous examples, except that the load is a distributed load of 4 k/ft.

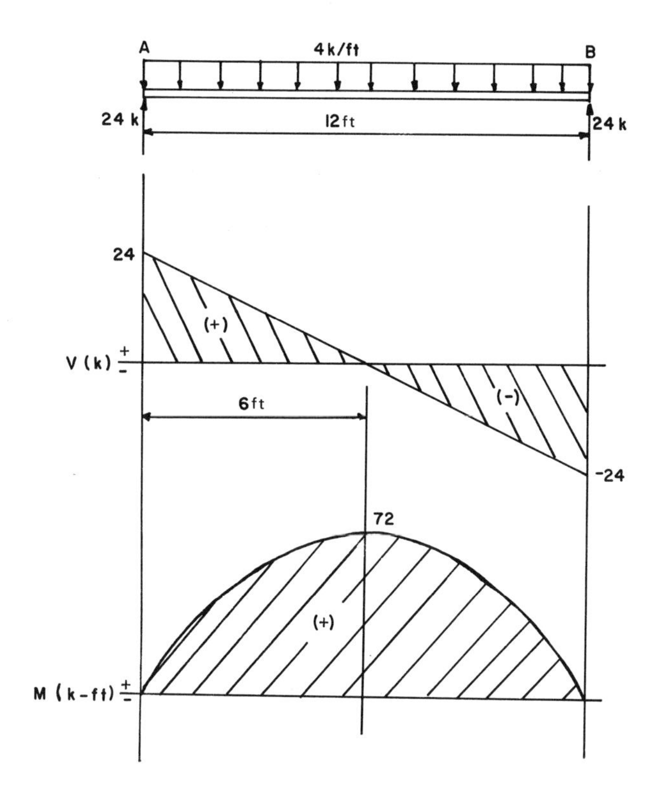

Fig. 4.4 Beam model and Loads for Example 6

$$\Sigma M_A = 0$$

$$4(12)(6) - R_B(12) = 0$$

$$R_B = 24 \text{ k}$$

$$\Sigma F_V = 0$$

$$R_A + R_B - 4(12) = 0$$

$$R_A = 24 \text{ K}$$

The shear is determined from the external loading. The reaction at the left end of the beam is equal to 24 kips (k). The distributed load of 4 k/ft is downward and therefore decreases the shear at a rate of a k/ft length. At total change in shear is

$$\Delta V = -4(14) = -48 \text{ k}$$

$$V_B = V_A + \Delta V = 24 - 48 = -24 \text{ k}$$

The reaction at B is 24 k upward for a net shear of zero at the end of the beam.

The moment at A is zero. The moment at midspan is determined by the change in moment between A and midspan or the area under the shear diagram.

$$\Delta M = \frac{1}{2}(24)(6) = 72 \text{ ft} \cdot \text{k}$$

$$M_{\mathbb{C}} = M_A + \Delta M = 0 + 72 = 72 \text{ ft} \cdot \text{k}$$

The moment at B is

$$\Delta M = \frac{1}{2}(-24)6 = -72 \text{ ft} \cdot \text{k}$$

$$M_B = M_{\mathbb{C}} + \Delta M = 72 - 72 = 0$$

4. Example 7: Truck on Bridge—Shear and Moment Diagrams

Draw the shear and moment diagrams for the two-span bridge shown in Fig. 4.5, load with the standard HS truck.

The reactions for the left span are computed as follows:

$$\Sigma M_A = 0$$

$$32(3) + 32(19) + 0.9(21)(10.5) - 21R_D = 0$$

$$R_D = 73 \text{ k}$$

$$\Sigma F_V = 0$$

$$R_A + R_D - 32 - 32 - 0.9(21) = 0$$

$$R_A = 39.9 \text{ k}$$

The reactions for the right span are

$$\Sigma M_E = 0$$

$$0.9(21)(10.5) + 8(12) - R_G(21) = 0$$

$$R_G = 14 \text{ k}$$

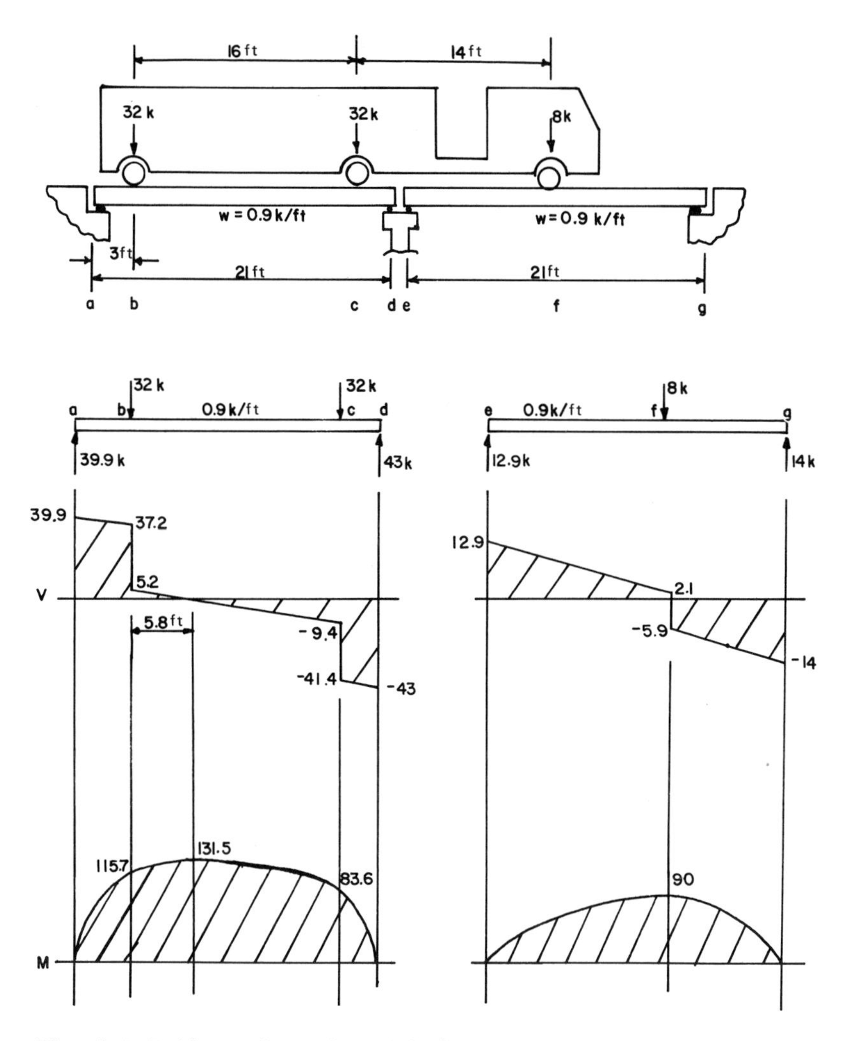

Fig. 4.5 Bridge and Truck Models for Example 7

$$\Sigma F_V = 0$$

$$R_E + R_G - 0.9(21) - 8 = 0$$

$$R_E = 12.9 \text{ k}$$

The shear for each span is determined from the loads as in the previous examples. The applied loads for this example are both distributed, representing the dead load, and concentrated, representing the live load of the design vehicle. On the left span, the reaction at point A is upward and produces a positive shear of 39.9 k. The distributed load decreases the shear at a rate of 0.9 k/ft to a value of 37.2 k just to the left of the load at point B. The concentrated load further decreases the shear to 5.2 k just to the right of point B. The distributed load continues to decrease shear to a -9.4 k at the left of C and the concentrated load lowers the shear to -41.4 k. The distributed load decreases shear to -43 k at point D which is equal to the reaction at that point. The shear diagram in the right is developed in a similar manner.

The bending moment at point A is zero. The change in moment is equal to area of shear diagram, therefore the moment at B is computed as follows.

$$\Delta M = \frac{1}{2}(39.9 + 37.2)(3) = 115.7 \text{ ft} \cdot \text{k}$$

$$M_B = M_A + \Delta M = 0 + 115.7 = 115.7 \text{ ft} \cdot \text{k}$$

The maximum moment occurs at the zero shear point which occurs 5.8 ft from point B.

$$\Delta M = \frac{1}{2}(5.2)(5.8) = 15.8 \text{ ft} \cdot \text{k}$$

$$M_{MAX} = M_B + \Delta M = 115.7 + 15.8 = 131.5 \text{ ft} \cdot \text{k}$$

The negative shear area then produces a decrease in the bending moment.

$$\Delta M = \frac{1}{2}(-94)(10.2) = -47.9 \text{ ft} \cdot \text{k}$$

$$M_C = M_{MAX} + \Delta M = 131.5 - 47.9 = 83.6 \text{ ft} \cdot \text{k}$$

The change in moment from C to D should produce a moment of zero at point D if the law of equilibrium is to be satisfied.

$$\Delta M = \frac{1}{2}(-41.2 - 43)(2) = -84.2 \text{ ft} \cdot \text{k}$$

$$M_D = M_C + \Delta M = 83.6 - 84.2 = -0.6 \text{ ft} \cdot \text{k}$$

The value at D is not quite zero but this error is caused by rounding-off the numbers during computations. The moment diagram for the right span is determined in a similar manner.

Reference

1. American Association of State Highway and Transportation Officials, Manual for Maintenance Inspection of Bridges 74, Washington, D.C., 1974.

Chapter 5

REPORTING SYSTEM

I. Numbering Sequence

A. Element Numbers Along the Bridge

Any numbering system used must specify a method for determining the starting point. The numbering system that will be discussed here uses the direction-of-route to determine the starting point. Direction-of-route is the direction from the starting point of a highway or road to its end point. If a highway has mile markers then the direction-of-route is in the direction of increasing markers. For standardization, routes in the United States are generally considered to start in the west or the south and end in the east or the north; therefore, a route can be said to travel from west to east and from south to north. The most practical numbering system relates a bridge to its roadway. Such a system designates the end of the bridge that is crossed first by the route as the starting point for numbering.

If the road is north-south oriented and if the bridge on this road is north-south oriented, the route is considered to cross the southern end first; therefore, the bridge elements will be numbered from south to north.

If the road is east-west oriented and if the bridge is also east-west oriented, then the bridge elements will be numbered from west to east.

It is the general orientation of the road that determines where the route starts and, therefore, determines which end of the bridge is first crossed by the route. This end is the starting point for the number system.

B. Element Numbers Across the Bridge

For elements that are to be numbered across the bridge, stand facing in the direction of the route and number or letter the elements from your left to your right. For example the columns in a bent are labeled 2a, 2b, etc., from left to right.

II. Sequence of Inspection

A well-planned sequence will provide the inspector with a means of more efficiently utilizing his time at the bridge site, and will assist him in making a systematic inspection. Most important, a planned sequence should ensure a more thorough inspection.

III. The Inspector's Notebook

A notebook, or for that matter any type of record, should be able to communicate to other individuals an assessment of the conditions observed during the inspection of the bridge. The notebook should be prepared in advance for the following reasons:

To help plan the inspection

To take advantage of better working conditions when preparing the notebook

To save time during the inspection

A. The Purpose of Sketches

Since the bridge inspector will occasionally need to execute his own sketches he should have some basic drafting skills. While he is not expected to have the ability of a draftsman, he should be able to make an easily understood sketch of any part of a bridge.

The inspection notebook will normally contain numerous drawings and sketches and may take a great deal of time to prepare or complete, particularly if the inspector has to draw most of the diagrams and sketches himself. But, this time will be well spent since systematic preparation will help to avoid oversights during the inspection. And, in many cases, it will probably be possible to insert reproductions of selected portions of the bridge plans or polaroid photos into the notebook. Most sketching is done in the field in order to more clearly document any unusual condition. Such sketches should amplify narrative descriptions of conditions and assist the reviewing engineer in his understanding of inspection reports.

Once the bridge inspector has learned the function of the various elements of the bridge, he will be able to subdivide most bridges into component parts. There are a number of unique sketches that may be required for certain special or complex types of bridges. The new or inexperienced bridge inspector should be aware of these requirements, but should not be overly concerned since he will not be expected to know how to prepare every type of special sketch. Only an experienced inspector would be able to independently compile a complete list of the necessary drawings and sketches. Any inspector, however, should be aware of the reasons for special sketches. In those cases that demand special sketches, the inexperienced inspector can expect assistance and guidance from his supervisor. As an inspector gains experience and additional knowledge of the more critical functional areas of the bridge, he will be better able to determine what type of sketches will be needed. Sketch determination is based upon the situation and circumstance; it is not an arbitrary selection process based on an established list of standard sketches.

B. Suggested General Format

Notebooks are generally the most suitable reporting format for complex or unusual bridges. Because they are used for complex bridges, they must be organized systematically.

The left-hand page should have:

Name of the structure

Page number

Name of the member being inspected

Element of component number

Element or component name

Rating

Problem

Comment

The right-hand page will be used for sketches or drawings. The notebook should be arranged sectionally, in the order in which the structure was built, and each section should be arranged so as to proceed from the general to the specific.

C. Title Page

The title page should contain:

The name of the structure

The structure identification number

The road section identification number

The name of the crossing

The back of the title page should contain:

The person in charge of the party

The names of the members of the inspection party

The type of inspection

The dates on which the inspection was made

D. Overall Sketch

The first sketch should schematically portray the general layout of the bridge and should include:

Illustrations of the structure plan and elevation with proper data

Immediate area description

The stream or terrain obstacle layout

Major utilities

Any other pertinent details (vertical and horizontal sketches)

If all the data cannot be placed on a single, general layout sketch, then additional sketches should be made as necessary.

E. Index Page

Place the drawing on the right-hand side with the appropriate items numbered on the left-hand side.

F. Substructure

Place the drawing on the right-hand side with the appropriate items numbered on the left-hand side. Items to be numbered, if applicable, include:

Pilings

Footings

Vertical supports

Lateral bracing of members

Caps

G. Superstructure

Items to be numbered, if applicable, include:

Main supporting member

Bracing

Wind bracing

Bearing devices

Sketches should be made of each major bearing device and of any bearing that is not functioning properly. Items to be numbered, if applicable, include:

Main supporting members

Floor beams

Stringers

Diaphragms

H. Deck

Items to be numbered, if applicable, include:

Decks

Expansion joints

Curbs

Hand rails

Any other items necessary such as scuppers, weeps, etc.

I. Sketches

Special sketches may have to be prepared for critical areas of certain bridges.

Sketches of fenders, dolphins, and aerial navigational lighting will sometimes be required. Any special or unusual feature that needs clarification will require additional sketches.

It is a good practice to leave every third or fourth page blank for additional sketches that may be required.

Alignment should be checked. Structures that are extremely long or show evidence of misalignment should be surveyed.

Each substructure unit sketch should show two end elevation views, one on each side. And, each substructure unit sketch should show one side elevation view. Both end and side views should be large enough so that misalignment, if any, can be plotted on the appropriate sketches.

A plan or overhead view should be drawn so that horizontal misalignment if any can be plotted. An elevation or side view should be drawn so that vertical misalignment if any can be plotted on the sketch.

J. Other Items

Environmental features that should be noted for the record are as follows:

Terrain features

Stream profile

High water elevation

Spur dikes or other protection devices

Slope protection

Channel protection

Channel location where it passes under the structure

The alignment of the approach roadway

Underwater investigation plots should account for scour. The final section of the notebook should include comments on mechanical and electrical systems. Reports of specialists should be added if available.

IV. Standard Forms and Reports

A. General

On large or complex bridges, a notebook format should be used to record inspection results. However, for small or simple bridges, it may be more practical as well as more convenient to use your state's standard forms. When using the standard forms, adjectival evaluation entries and supporting comments may be recorded in much the same manner as in the inspector's notebook outlined above. If available, standard, prepared sketches should be attached to the inspection form, unless this would be considered inappropriate.

All bridge elements referred to should be clearly identified in accordance with practices and procedures of your state highway organization, and cited appropriately in the narrative. All items on a form may not always have to be completed. Prior to the inspection, it should be determined which items are not applicable to the structure or sign being inspected; these should be either lined out or otherwise indicated as not being applicable.

B. Overall Inspection Forms

A summary sheet is normally filled out and is used as a cover page for the inspection report. This summary sheet also serves as a source of overall inspection information. Usually the most pertinent information from the previous inspection of a particular bridge can be obtained from the summary sheet. Such information can be of considerable assistance to an inspector in comparing existing bridge conditions to those previously reported. Significantly, it allows him to audit what actions in the way of maintenance or repairs have taken place since the previous inspection.

Many of the standard inspection forms consist of a single sheet that applies to a specific element. This type of inspection form lists for inspection only those items related to that element. Usually, items other than those listed are also to be inspected. A listing of all items serves as a checklist and should preclude any oversight on the part of the inspector. Moreover, such a checklist should indicate to the inspector the tools and equipment he will need to adequately accomplish his inspection task. Before leaving for the bridge site, the inspector, in a final check, should make certain that he has the tools and equipment he will need during his inspection and whatever forms are required to record his inspection observations.

C. Accurate and Detailed Reporting

The bridge inspector, in order to be as objective as possible, should record exactly what he sees and measures; nothing more. The inspector may feel at times that he can resort to short cuts in his reporting; but if he does, he may not fulfill his obligation to the highway department or to the motoring public, which uses the structure.

At the bottom of the summary sheet there is an excellent place to make a quick summary of any problem areas encountered during an inspection. If an inspector feels that it will be helpful to those reviewing his report, he may elect to summarize briefly his inspection findings. Such summaries can be placed at the bottom of each sheet or on the reverse side of the sheet.

The inspector should recommend further investigation whenever he feels that he is incapable of making an accurate determination of any problem. This may be necessary in those cases that require more sophisticated equipment or when the inspector feels that he is not adequately trained for operating the equipment. The inspector should not be reluctant to ask for assistance.

Any need for immediate repairs should be ascertained by the inspector. He should be aware of the need to make recommendations for immediate repairs when existing conditions indicate that prompt attention is in the best interest of public safety, of the integrity of the bridge, and of cost effectiveness.

V. Rating of Bridges

A. Descriptive Ratings

As required, you will gain experience in rating bridge elements and components in accordance with four descriptive ratings agreed to by the respective states. These descriptive ratings are:

Good

Fair

Poor

Critical

It is to be emphasized that the descriptive ratings will be assigned to individual bridge elements, components, and conditions. That is, the deck could be rated as good, fair, poor, or critical. Likewise, the condition of the paint could be rated as good, fair, poor, or critical. The criteria governing the rating to be assigned is "how well is the element, component, or item fulfilling the function for which it was intended." Therefore, while the AASHTO numerical rating refers to the condition of broad areas such as superstructure and substructure, the inspector, under the descriptive rating system, need concern himself only with rating the condition of individual elements such as abutments, piers, bents, bearings, beams, decks, and so on. The inspector's rating of the individual elements will form the basis whereby the structure may be evaluated by higher authorities.

1. Good A rating of "good" corresponds to AASHTO condition ratings 8 and 9. That is, the element is new or in good condition with no repairs necessary.

2. <u>Fair</u> A rating of "fair" corresponds to an AASHTO rating of 5, 6, or 7. That is, the element is either a minor or major element in which the potential exists for minor or major maintenance or minor rehabilitation.

3. <u>Poor</u> A rating of "poor" corresponds to the AASHTO rating of 3 or 4. That is, for the element concerned, the potential exists for major rehabilitation. Or the element presently requires repair or rehabilitation to perform its intended function satisfactorily.

4. <u>Critical</u> A rating of "critical" corresponds to the AASHTO rating of 0, 1, or 2. That is, the element concerned is not performing the function for which it was intended. If a pile is not providing the support it is intended to supply, then it is in critical condition. The bridge should be closed.

B. Review

Descriptive ratings are assigned to individual elements, components, and conditions. The rating assigned relates to the effectiveness of the element, component, or conditions of a bridge structure in performing the job for which it was intended. For example, bridge drainage is rated on the basis of how well the drainage system performs its function of providing a means for water to be drained away from the bridge structure. Piers and bents are rated on how well they perform their function of supporting the superstructure.

C. Amplifying Information

Up until this point a code of descriptive ratings used by the bridge inspector to evaluate observed conditions of deterioration in bridge elements and components has been enumerated. These ratings have been correlated to the AASHTO numerical ratings of 0 through 9, and examples of good, fair, poor, and critical have been discussed. The ability of a bridge inspector to rate observed conditions is certainly important. However, there are other equally important facets of the bridge inspection job.

1. Photographs

In documenting deterioration, photographs of the affected area should be taken. This provides a permanent record of the condition at the time of inspection. The use of a color polaroid camera is desirable since it provides the inspector with on-the-spot assurance that he has photographed the observed condition in its true perspective. Then too, the picture can be annotated with amplifying information such as the rating assigned, the type of deterioration, the extent of deterioration (such as the depth to which corrosion has penetrated a member), the identity and the location of the member. However, a 35-mm photograph is also required for permanent records. When taking a photograph it is desirable to include a strip of tape measure in the picture for scale and reference.

2. Sketches

Even if a polaroid camera is used, a sketch of the element and the deteriorated condition is needed.

3. Measurements

Measurements of defects are obviously important. In addition, other measurements such as those associated with bearing moment, joint width, and the like are required to determine whether these elements are functioning properly. What to measure, how to measure, and what to record will be explained throughout the text.

4. Narrative Description and Summary

Only the inspector is qualified to describe in narrative form what he or she has observed. His or her comments should be clear and short, accurate and complete. The inspector should also complete a written summary of the observed conditions, listing the more critical defects first. The report on the individual elements, components, and conditions must be accurate and complete enough for others to evaluate the structural integrity of the bridge and to know when corrective action is required.

5. Recommendations

The inspector's job is not done until he or she has completed the report and made recommendations for corrective action, where required. The inspector should list repairs which may be required in the order in which they should be performed; i.e. critical repairs should be listed first, followed by repairs to elements in poor condition, and then repairs to elements in fair conditions.

If the inspector has any doubt about his observations, he or she should not hesitate to recommend that the condition in question be examined by more sophisticated techniques or that a more qualified individual examine the situation. Finally, the inspector must submit his or her report and recommendations to proper authority. Only when this has been done, and the inspector is sure that the report and recommendations are understood, is his or her job complete.

Chapter 6

SUPERSTRUCTURE INSPECTION

I. Decks

Bridge decks probably suffer more wear and deterioration than other bridge elements because of the direct effects of vehicular traffic, weather, and the application of chemicals and abrasives to the deck surface. Inspection procedures for decks are mostly determined by the material the deck is made of and are much the same as procedures for inspecting any other bridge elements constructed of the same materials. For example, assessing deterioration of timber decking due to decay is similar to inspection for decay of other timber bridge members. However, the abrasion and deterioration due to traffic contact are unique to deck inspection. The functional characteristics of decking and other features of the superstructure will be related in this chapter to the types of deterioration normally encountered when inspecting bridge decks.

Bridge decks serve other purposes than merely to provide a roadway over which traffic can move. Decks, like floor systems, also distribute traffic or vehicular loads and their own weight to other bridge elements. Depending upon the bridge design and construction materials used, these wheel loads may be:

Transmitted to the floor system beams and stringers

Transmitted to the main supporting members

Transmitted directly to the pier caps or abutments in the case of concrete slab bridges

Bridge decks which are cast or connected to the floor system beams, guides, or stringers also may act as the upper flanges of those members. Such design is termed composite construction and assists those supporting members in resisting the compressive forces normally found in these flange areas. In this instance, the composite area within the deck might be considered as acting to support deck loads as well as acting to distribute deck loads to other bridge support members.

Deck materials may be timber, concrete, or steel, or a combination of these materials. The physical and mechanical properties of these materials, described earlier, help the inspector determine the types of deterioration which may be encountered during on-site inspection of decks. Constant exposure to atmosphere conditions makes weathering a significant cause of deck deterioration.

Vehicular traffic across the deck produces impact, rolling, sliding, and other damaging effects on the deck surface. In some instances a wearing surface is applied to the deck surface, either to better resist traffic wear and improve skid resistance, protect the surface of the deck itself, or both. Such surfacing material generally consists of a coarse aggregate and/or binder substance, which binds the aggregate and also binds the surfacing to the deck.

Concrete decks that are cast in place occasionally are supported from beneath by prefabricated stay-in-place (S-I-P) forms. These forms are metal sheets permanently installed above or within the floor system. After the concrete has cured, these forms (as the name indicates) remain in place as permanent nonworking members of the bridge. They are subject to those types of deterioration that are typical for the metal and common to other, similar metals.

Concrete decks also contain steel reinforcement, or rebars as they are commonly called. Exposure of these internal elements to moisture or corrosive agents will cause subsequent deterioration of the steel and lead to further deterioration of the concrete.

II. Floor Systems

The purpose of floor systems is to support the bridge deck or slab and transmit the weight of the deck and deck loads to the main supporting members of the bridge itself. These purposes are served by either the floor beams, the stringers, the main supporting members such as girders, or by combinations of all three. The length and width of the span and the type of construction materials used to bridge the distance largely determines the composition and configuration of the floor system.

Floor systems have two principal components, floor beams and stringers. Floor beams are the transverse members which extend perpendicular to the traffic flow and which support the stringers or deck slab and also transfer the deck loads to the main supporting members of the bridge—beams, girders, or trusses. In some instances the pier caps may also function as floor beams where the deck slab or stringers rest directly on the cap.

Stringers run longitudinally and parallel with the roadway. They are spaced across the length of the floor beams and either rest upon the floor beams or are connected by angles to the web of the floor beam. The variety of materials—timber, concrete, and steel, plus the variety of cross sections and designs used for floor systems dictate that care be exercised during inspection in order to describe accurately conditions of specific floor system members.

There are basically three types of floor systems, which are described in Bridge Inspector's Training Manual 70 (DOT Manual) [1].

Type I: The deck is supported by the main supporting members of primary stringers. This type of floor system is quite commonly found on timber stringer bridges.

Type II: The deck loads are carried by both stringers and floor beams and are transferred to the main supporting members. In the case of girder bridges the floor beams are attached to the web of the main girders. The floor beams in truss bridges are connected to the trusses at the upper or lower panel points depending upon whether the bridge is a deck, or a thru-truss-type.

Type III: The deck is simply supported at each end by the floor beams or caps. This type may be encountered on long, narrow spans where the transverse or lateral strength of the slab is sufficient to resist buckling or transverse deflection.

III. Beams and Girders

While there is a fine distinction between a beam and a girder, the terms normally refer to bridge members used to resist flexure or bending resulting from loading perpendicular to the member. The distinction between a beam and a girder lies in the fact that a girder is normally a primary supporting member in the bridge and may receive loads from beams. Therefore, any large beam could also be classified as a girder. In the case of timber members, they are normally referred to as timber beams. A rolled-steel wide flange section is also called a beam.

A. Loads

Beams and girders, when acting as the main supporting members in the superstructure, transmit the loads supported to the substructure. These loads are normally differentiated as dead, live, and impact during the design of bridges. For the purposes here, in regard to beams and girders, dead loads are considered to be distributed at a uniform weight over the length of the span. Live and impact loads due to the axle leads of vehicles are considered to act at a specific point or points representing the wheel contact points on the beam or girder.

Stresses produced, such as tension, compression, or shear, occur at specific locations in direct relation to the location of the load and the location of the substructure support points for the beams. The construction material must be adequate to carry the stresses produced. The number and location of substructure supports for a beam will also indicate the location and the types of stress produced in the beam by various loadings. For example, beams simply supported by a support at each end are stressed differently than beams that are continuous over an intermediate support.

B. Types of Spans

There are generally three types of spans for which beams and girders are used. The most common is the simple span. The simple span entails a support at each end. The significant stresses normally will occur at the midpoint between supports for bending and at the end supports for shearing.

Continuous spans extend over one or more intermediate supports between the end supports of the span. As the span is subjected to live loadings, the areas near the intermediate supports (normally at the one-quarter point) and directly under the loads are subjected to stress reversals.

In discussing cantilever spans it is sufficient for our purposes to note that the stresses produced in the beam at the support point are similar to those produced at the intermediate supports for continuous beams. These stresses are of primary interest.

C. Cross Sections

The cross section of beams and girders is of main interest to the inspector in that its shape or configuration plus cross section area plus the type of material gives a fair indication of the type of the stresses for which it was designed. Typical cross sections are rectangles (solid or hollow), T, I, and wide flange (W). Wide flange beams are also used as stringers or main carrying members. Solid rectangular sections are normally timber whereas the hollow rectangular section may be reinforced or prestressed concrete. The horizontal portions of the cross section are termed flanges and the vertical portions are referred to as the web. The flanges usually carry the tension and compression stresses as a result of bending, while the web resists shear. When the required depth of the steel cross section cannot be satisfied by a mill-rolled section, a built-up beam or a plate girder is used.

Plate girders are made of steel and, as the name implies, comprise the web plate and flanges fabricated from angles, channels, plates, or a combination of all three. These are fastened together by rivets, bolts, or welds, or various combinations thereof. Again, the shape and area of the cross section are designed to carry specific design loads. Stiffeners may be found along the web to resist buckling, particularly at the bearing points over or near the girder supports. Stiffeners are usually angles or plates that are riveted or welded to the web.

D. Connections

Timber beams are usually connected by means of bolts. Washers beneath the nuts and bolt heads prevent crushing as the connection is drawn up. Since many timber beams also serve as stringers, there may be instances in which their thickness will permit the use of spikes or drift pins as the beams lap at the support point. Metal rings (split or toothed) and other connector devices are used between the joined timbers to help resist movement in the joint. Lateral movement and buckling may be resisted by bracing between the beams.

Concrete beams may be precast or cast in place. Concrete beams may be reinforced or prestressed. In simple spans they are supported at each end by the abutments or pier caps. Continuous concrete beams are usually cast in place. Where stiffening is required, diaphragms may be cast monolithically with the beams, and the extension of diaphragm reinforcement steel into the beams provides the connection. Similarly, the deck may also be cast monolithically with the beams, and extension of the beam stirrups into the slab provides the connection. Pins or dowels set in the cap or bridge seat may also be used; they would fit into holes at the ends of precast beams.

Connections used for steel beams and girders are usually angles or plates; again, either riveted, bolted, or welded at the flange or web portions of the members connected. Where beams or girders are spliced, splice plates are employed at the joint. During erection, small angles (clip angles, angle seats, or shelf angles) may be used to temporarily hold the beams or members in place until the entire connection can be completed. While these angles do not always carry the connection load, they do contribute to the strength of the entire connection and should be inspected in the same manner as the connection itself. Conversely, connections or clip angles used at the end of stringers or floor beams may carry the full load.

IV. Trusses

A truss is a structural frame composed of individual straight members connected to form a series of triangles. A truss, rather than girders, is usually used for longer spans. And, like a beam and girder, it may be simply supported at the ends, it may be cantilevered, or it may be continuous over one or more intermediate supports. When the truss is considered as a single plane, that is, lying in one plane and resting on the bridge bearings, it is acting as a main supporting member. Trusses used for other purposes, such as lateral bracing, will be covered accordingly. Thus, the truss as a structural member does provide great versatility.

Types of truss bridges may be either through or deck, depending upon the location of the floor system or deck. When sufficient height (or depth) of truss is not provided to permit overhead bracing the truss is referred to as a pony, or half-through truss.

A. Truss Terminology

In Fig. 6.1, the bridge consists of two vertical trusses which carry the loads transmitted by the floor beams. . The floor beams support longitudinal stringers upon which rests the deck. Two horizontal trusses in the plane of the top and bottom chord carry the wind loads. Secondary trusses in the plane of the end posts (portals) and in the plane of the intermediate posts are called sway bracing. The elements of the vertical trusses are vertical posts, main ties, counters, top chord, bottom chord, and end posts.

B. Truss Patterns

There is a great variety of geometrical truss patterns. In most instances a particular design characteristic has been named for the original designer. Several of the more common truss patterns are shown in Fig. 6.2. The

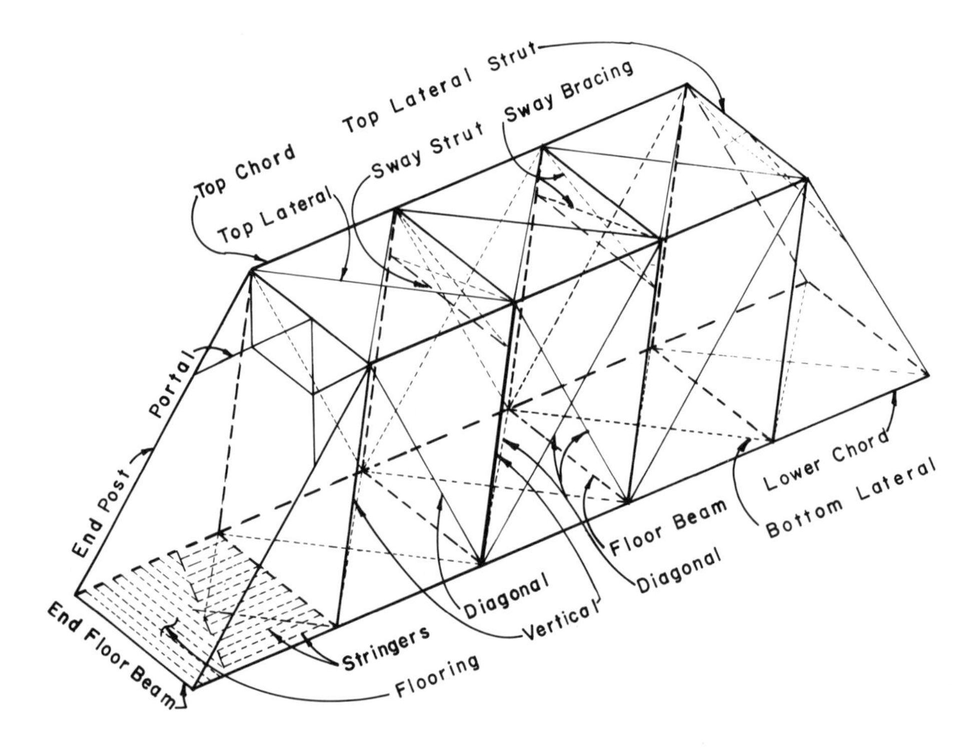

Fig. 6.1 Components of a Typical Thru Truss

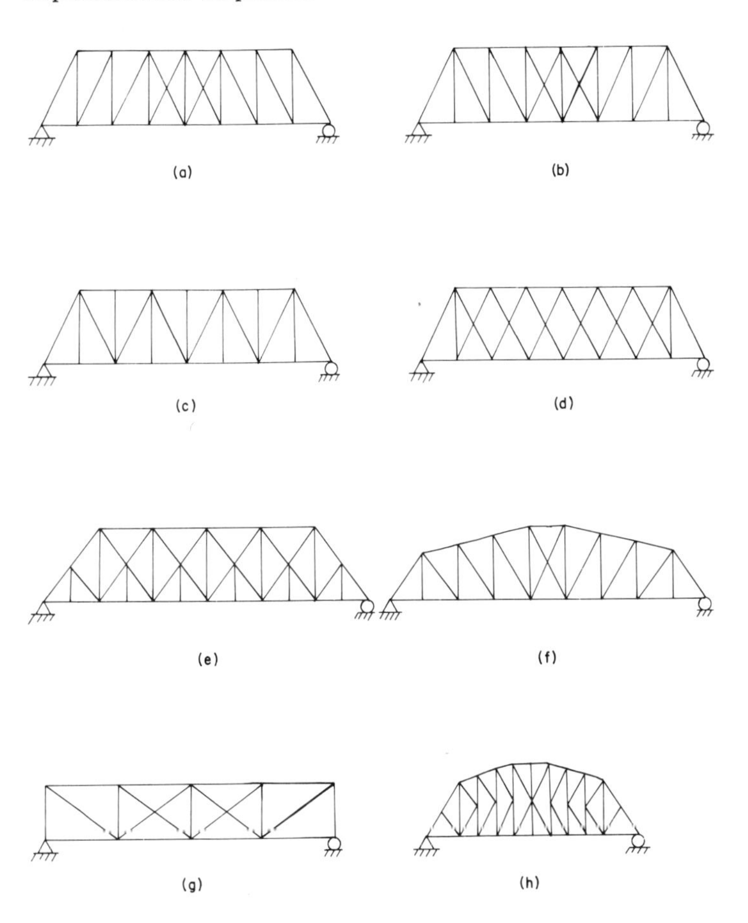

Fig. 6.2 Various Truss Bridge Configurations

location and directional orientation of individual truss members in each truss give some indication to the bridge inspector of the type of stress that is normally found in that member. The Howe truss (a) has its vertical web members in tension and its inclined members in compression. The Pratt truss (b) has its vertical members in compression, while its diagonal members are in tension. A Warren truss (c) is a combination of isosoles triangles. The tension and compression members are determined by location and configuration. A Quadrangular Warren truss (d) has alternating compression and tension diagonals. The Baltimore truss (e) is a Pratt truss truss with parallel chords in which the main panels have been subdivided by an auxilliary framework. The Camel Back (f) is a Pratt truss with a variable height for the upper chord. A deck truss (g) may be any one of several truss patterns but the deck rests on the top chord rather than on the bottom chord (as with the more common through trusses). The K truss (h) refers to a particular K pattern of the diagonal members of the truss.

C. Stresses

To simplify the understanding of stresses that occur in trusses and truss members, basic assumptions are usually made during design and stress analyses of trusses. These assumptions are generally stated as:

> Members of the truss and forces (loads) that act on the truss lie in the same plane.
>
> Forces upon truss members act along the longitudinal axis of the member, producing either tension or compression stresses in the member.
>
> Individual members in the truss are rigid and connected at their ends by frictionless pins.
>
> Loads and reactions on the truss act only on the pins (at the panel points).
>
> The weight of the members is neglected initially.

These assumptions vary in differing degree for actual truss bridges. However, they do assist in understanding the basic factors which influence what the bridge inspector looks for, where he or she looks, and why he or she should inspect for that particular condition.

In general, the upper chord of a simple-span truss is placed in compression and is usually termed a compression member as a whole or in reference to each segment. Conversely the lower chord is placed in tension and is termed a tension member. Web members (verticals, diagonals, and counters) can be either compression or tension members, depending upon the type of truss and their location in the truss.

The loads that act on trusses are the same as those acting on beams and girders. The dead load for each truss consists of the weight of the truss and one-half of the weight of the deck and floor system acting uniformly along the span of the truss. It is divided evenly among the panel points between the supports. Live loads (including impact loads) are the axle loads determined by AASHTO for a vehicle or vehicles expected to transverse the bridge. In the truss design, these live loads are moved along the truss to produce the maximum stress in each member, and the truss members are designed to accommodate anticipated or planned stresses. For this purpose, it is adequate to understand only that these loads act vertically at the panel points where the floor systems is attached.

Trusses may be supported at the ends by bearings. The bearings at one end are usually of the roller type so that the reaction at this support is in a vertical direction.

D. Cross Sections

The ability of truss members to resist tension or compression stress is determined by the available cross-sectional material and its configuration or shape. The versatility of steel fabrication methods permits a variety of steel truss member configurations that one will encounter during on-site truss bridge inspections. In many instances the state of the truss member, tension or compression, may be determined without a stress analysis but with simply a visual observation of the truss pattern.

Tension members include: rods, eye-bars, beams, channels, wide flange sections, angles, beams, and channels connected by plates.

Compression members include: T beams, channels, wide flange sections, angles, beams, or channels latticed or plated together.

E. Connections

Where truss members are built up or latticed, connections are made by a riveting or welding process. In the case of riveted tension members, the total available effective cross section is reduced by the number and size of the rivet holes required. In welded members, depending upon the type of weld, full strength of the cross section can be developed if a sufficient amount, length, and type of weld material are provided.

Where rivets are used for fastening, stress in the member tends to shear the rivet. Shearing can be on a single plane or two planes depending upon the number of parts that the rivets are intended to hold together as well as the direction of stress in those parts. Conversely, rivets exert a bearing stress on the materials through which they pass and will have a tendency to tear through the material that they hold together or crush the material in bearing.

V. Preparation for Inspection

The routine inspection of bridges should follow the procedures outlined in _Bridge Inspection Training Manual 70_ [1], the AASHTO standards [2], the _Recording and Coding Guide_ [4], and the Department of Transportation _Manual on Bridge Maintenance Inspection_ [3]. The routine inspections are required at least once every two years and will consume most of an inspector's time.

Interim inspections are needed for bridges that are deficient. This deficiency may consist of weight limitations and/or geometrical deficiencies or structural deficiencies. Such inspections should be scheduled as often as necessary to ensure public safety. Interim inspections are often required for bridges that are damaged by accidents, fire, or for other reasons.

A. Scheduling

A schedule should be developed for all types of inspections. This schedule should be followed as closely as possible so that all structures will be inspected within the required time span. Scheduling should be far enough in advance so that time is available for coordination of manpower, equipment, and specialist personnel. Weather, both seasonal and forecast, stream levels, and seasonal traffic loads must be taken into consideration during scheduling. Enough latitude must be allowed so that manpower will be available for special inspections or unforeseen problems.

If the people who schedule the inspections were able to predict the future, their job would be easy. Problems arise from not being able to employ the number of people or have the amount of equipment adequate for all circumstances. Usually inspections go fairly smoothly until a problem arises that demands immediate attention, and then everything gets behind schedule.

Completing an inspection within a given time frame is difficult even when only one or two people are involved. The addition of more people and support equipment adds to the complexity of the inspection.

Careful planning is essential for a complete and efficient inspection. During this phase, several items should be considered. The inspector should note when on-site time has been scheduled, and the amount of time allowed for preparation of the inspection, the nature of the inspection itself, and the type of report to be written after the inspection. It is important for the inspector to predetermine the type of inspection so he or she may plan his or her sequence for the inspection and plan for the type of equipment needed.

The crew chief should determine the resources needed and consult an inspection check list. On a small bridge it may be that only one or two people, a step ladder, and a few other items are required. However, even on small birdges the inspectors should make sure that their personal equipment is ready for use. For example, shoes should be in good shape if climbing is required, they should have adequate warm clothing in cold weather, and they should also have sufficient reflective clothing. Sufficient signing is always necessary.

On a large bridge, which may take a week or longer to inspect, the following personnel and equipment may be required: a supervisor, many inspectors, and specialized people such as divers, heavy equipment operators, boat operators, etc. Normal inspection equipment plus a boat, mechanical lifts, ladders, scaffolding, electronic measuring equipment, etc.

For most inspections all needed manpower, equipment, materials, and instruments are readily available to the inspecting agency. There are times, as in the case of the large bridge mentioned, when special resources must be obtained from outside sources. Special resources may include mechanical, electrical, hydraulic, or underwater inspection specialists, or special instruments such as ultrasonic measuring devices, electronic depth finders and others, or special equipment such as additional scaffolding, cherry pickers, and snoopers.

B. Data Study

The inspector should review the as-built plans if they are available or the design plans. The review should include the following:

The stress sheets are used to note which members are in compression and which are in tension. Usually compressive stresses are marked minus and tensile stresses are marked plus.

The stream's channel.

The flood plane.

Utilities on the right-of-way.

The right-of-way itself.

Soil and foundation information.

Stream profile and stream protection devices.

Stream elevation at normal and high-water levels.

If the stream is used for water traffic, the navigation devices used should be noted along with the bridge protection devices.

Special attention should be given to complex areas that may cause problems.

Identification by the inspector of critical areas.

Careful review of the entire bridge plan.

A study should be made of all histroical data related to the structure including the following:

The most recent inventory report

The latest traffic data

Prior inspection reports

Reports of repairs and reconstruction work

When a sequence of inspection is formulated, use should be made of all previously collected information. Normally the inspection sequence will follow sequence of construction. Either a notebook or a standard form, or both, should be prepared and used prior to and during an inspection.

Coordination of resources is very important. If two people are involved and they have their own equipment, then the only major coordination problem is one of timing, seeing that they both are at work at the scheduled time and arrive at the bridge site together. In the coordination of many individuals and equipment from different agencies, difficulties arise even under ideal conditions. So imagine what it is like under simply normal conditions. Large complex bridges and bridge culverts are at the two extremes, with most bridges somewhere in between.

With the purpose and function of the primary components of a bridge in mind and the inspection preparation and sequence determined, the inspector is ready to begin the superstructure inspection. The inspection procedure includes three areas of activity: a visual inspection, a mechanical inspection, and appropriate measurements.

VI. Inspection Procedure

A. Decks

The roadway profile over decks should be checked for abrupt changes in elevation, which could indicate foundation movement, failure of supports, or other adverse conditions. A profile should be taken over the centerline and each gutterline using surveying instruments such as levels and rods and then compared with the original profiles, if they are available. These profiles should be recorded in the inspection report for all long spans and may also be recorded for shorter spans.

1. Timber Decks

Timber decks are normally referred to as decking or timber flooring and the term is limited to the roadway portion, which receives vehicular loads. The term should not be confused with the floor beams or stringers, which support the flooring and which are referred to collectively as the floor system. The decking is usually placed across the bridge, perpendicular to the flow of traffic. In most cases it is fastened to the floor system by nails or spikes.

The timber itself should be inspected for the normal indications of timber deterioration such as decay, splitting, or cracking. Where the timber itself provides the wearing surface it should be checked for traffic wear and abrasion. Timber decks should also be checked for slipperiness, especially when wet.

A wearing surface of bituminuous and aggregate mixture is often applied to timber decking. The condition of this wearing surface should also be checked for evidence of deterioration such as cracking, holes, or lack of adherence to the decking which may indicate deterioration of the decking beneath.

Timber decks should also be inspected from the underside for material deterioration and looseness of the decking. Excessive deflection of the decking under moving traffic loads should be recorded. This deflection may give an indication of the cause of wearing surface deterioration.

Examples of decks in various conditions are shown in Fig. 6.3. Fig. 6.3a shows a deck in good condition; the timber has not been damaged nor are there any signs of deterioration. Drainage from the surface is good. Fig. 6.3b shows a deck in fair condition. The timber surface is showing signs of wear and minor damage but the carrying capacity has not been significantly reduced. Drainage remains good. Fig. 6.3c and 6.3d show a deck in poor condition. The surface is badly worn with damage and deterioration evident. The capacity of this deck would be reduced significantly. Drainage would be slow on the surfaces shown. The timber deck shown in Fig. 6.3e is in critical condition. The ends of the decking planks show extensive deterioration. The effective cross section of the decking has been reduced to such an extent that the load carrying capacity must be very low.

2. Steel Decks

Steel decks may be solid or gridded. Steel decks are quite vulnerable to corrosion since they are exposed constantly to the weather and to continual traffic abrasion. Where there is evidence of corrosion the inspector should check for any loss of section. Measurements may be made with calipers to determine the reduction in material section. Deck fastening and connections such as welds or clips should be checked for cracks as well as for rusting of the connection.

Steel decks can be worn smooth by traffic abrasion. The deck surface should be checked for slipperiness, particularly when wet. Some steel decks are fabricated with an irregular riding surface or are modified to include small steel studding on the surface to improve traction and to counteract possible slipperiness.

3. Concrete Decks

The plastic characteristic of wet concrete that permits casting in various shapes and sizes has provided the bridge designer and the bridge builder with a variety of construction possibilities. The construction versatility of concrete also presents to the bridge inspector the same variety of items to

Fig. 6.3a Example of a Timber Deck in Good Condition

Fig. 6.3b Timber Bridge Deck in Fair Condition

Fig. 6.3c (left) Timber Deck in Poor Condition; 6.3d (right) Timber Deck Showing Wear–Poor Condition Rating

Fig. 6.3e Timber Decking Planks Showing Extensive Deterioration—Critical Condition

look for during concrete deck inspection. In general, concrete-deck inspection items can be classified as scaling, spalling, or cracking with an additional group of kind of effects caused by wear. In all instances it is helpful to the inspector if he or she has available the previous bridge inspection report so that the progression of deterioration of the concrete deck can be determined, thus providing a more meaningful report.

The location and extent or area of the particular type of deterioration should be both described and measured with a tape (for the record) in the inspection report. When the location and type of deterioration indicate that damage is due to other bridge elements, such as at expansion joints that have spalled edges because they are filled with incompressible material, those elements should also be checked for faulty design, construction, or conditions.

The wearing surface, if present as a separate layer, should be inspected for soundness and firm bonding or adherence to the deck. Where defects such as cracks or potholes are observed in the wearing surface, the cause may be deterioration of the deck beneath. Removal of some of the wearing surface material or checking the underside of the deck slab may disclose the cause. Those decks where the traffic rides directly on the concrete slab should be checked for wearing away of the cement paste and the finer aggregate, which results in larger aggregates becoming exposed and polished. Such conditions may increase the slipperiness of the deck when wet.

Deterioration of S-I-P forms should be checked for evidence of cracking in the deck slab above. Loss of the form or form material itself is not important since it does not contribute to the functional capability of the bridge. However, it does serve as a good indicator of a defect above, either in the deck or deck joints, which is permitting moisture or corrosive agents to filter through to the form surface.

Corroded reinforcing steel, either visible or indicated by straining, should be recorded. Such evidence indicates loss of strength within the concrete due to cracking and loss of bond between the concrete and the reinforcement itself.

B. Floor Systems

Loads received from the deck or deck slab should be applied along the total upper surface of the stringers and/or the floor beams. In the case of concrete bridges, the deck slab may be cast in place and tied directly to the stringers or floor beams by means of reinforcing rods. Certain steel beam or steel stringer bridges may have the deck slab cast directly on or around the upper steel flange. The slab is firmly attached by means of embedding the shear connectors or studs which were set in the flange of the steel beam into the lower portion of the deck slab. Similarly, many concrete T beam and box girder bridges utilize the upper surfaces or flanges themselves as the deck slab.

Where the deck merely rests upon the floor system, the adjoining surfaces are subject to abrasion and deterioration due to corrosion and possible movement, thereby affecting the even distribution of the deck loads across the floor system supporting surfaces.

Transferral of the deck loads by the floor system involves two distinct functional aspects. The first functional aspect involves the floor system member's ability to properly function as a beam in resisting the loads imposed by the deck. The stresses to be considered are compression along the upper flange area, tension along the lower flange area, and those stresses that may produce web crippling or transverse buckling between the end supports. At the ends, in the case of stringers or floor beams resting upon other members, crushing or vertical shear may occur near the bearing point.

The second functional aspect concerns the end connections where the stringers are joined to the floor beams or the floor beams are joined to the bridge girders or bridge trusses. The main connection angles or plates are riveted, bolted, or welded and should provide a fully developed joint.

Where seat angles are also used at the connection, they should be in firm contact with both joined members and provide additional support at the connection.

Bracing members within the floor system also contributes to the transmittal of deck loads to the main supporting members and provides stiffness as well to the floor system members and to the bridge as a whole. The stiffening effect of diaphragms on beams or stringers is discussed in the section on bracing, p. 100. These diaphragms prevent canting or transverse buckling of the stringers. On deck truss bridges, e.g., the lateral cross bracing truss may utilize the upper strut as a floor beam within the floor system. The remaining cross frame truss members contribute to the distribution of the deck loads applied to the truss at the upper chord panel point.

Where lateral bracing is attached by plates to the main supporting members, such as the main girders, common design practice includes connection to the floor beams as well. The angular loading effect of this connection to the floor beam must be included during inspection.

Floor beams and stringers may be made of timber, concrete, or steel. Therefore, the types of deterioration common to those materials must be considered during the inspection of floor system members. Similarly, the relative strengths of these members and their ability to withstand the types of stresses produced by various loading should also be considered in relation to their material composition.

Just as in other bridge members, there are certain conditions within each functional area that the inspector must look for in order to properly evaluate the floor system. The inspection should be complete if each functional area is examined for evidence of adverse conditions for each inspection item.

1. Timber

Timber floor system members should be inspected along the deck-bearing surface to ensure that the lower surface of the deck bears uniformly without crushing. Conversely, the timber stringers should be inspected for evidence of decay or deterioration that detracts from their ability to support the deck loads. These conditions may exist particularly where the individual timber-decking planks are not secure to the floor system or permit deck surface materials or drainage to filter through between planks.

Timber floor system members should be inspected at their support plates to ensure that there is adequate bearing area on the support and that no crushing has occurred.

Sighting along the length of the stringers should disclose lateral buckling, excessive sagging, or canting. If such conditions are detected the inspector should indicate whether they are of a permanent nature or occur primarily during vehicle passage. The report should also indicate partial weakness or failure.

The support areas should be checked for accumulations of debris and evidence of moisture or decay.

2. Reinforced Concrete

Reinforced concrete floor systems should be inspected as any concrete beam. However, concrete construction practices may pose certain problems in differentiating individual components for specific inspection needs. For example, where the deck slab is cast to include the upper portion of reinforced concrete beams, the exact line or bearing point at which the deck rests on the beam is a rather academic point. On the other hand, deck slabs resting on prestressed concrete beams should be inspected along the upper surface to ensure that there is good contact between the two and that no spalling is occurring in either the slab or the beam.

Cross bracing between concrete stringers such as diaphragms should be inspected to ensure full contact at their ends and that no spalling or cracking is present.

As with steel or timber beams, sighting along the axis of concrete beams will disclose excessive sag or canting. The lower flange area should be inspected for cracking, which may indicate overstressing in the concrete stringers.

3. Steel

Steel floor beams and stringers should be inspected at all connections. Steel floor system connections are particularly vulnerable to corrosion due to their exposure to moisture and chemical agents draining off the roadway. Connections tend to collect dirt and debris which in turn retain moisture and cause further deterioration.

The same corrosive condition may exist along the upper flanges that support the deck slab. Built-up floor beams must be carefully inspected for corrosion or rusting between the individual parts, such as at angles and where the angles join the web plate.

Floor system connections should be inspected for tightness. Loosening may be caused by bending, shock, or vibration. Missing or loose rivets or bolts should be recorded as should any movement of the plates or angles used in the connection.

Stringers and floor beams should be inspected for cracks in the web areas along the flanges, at the connections, and at the bearing ends over supports. Moving vehicle loads transmitted by the floor system can create considerable vibration within the floor system members, and such cracking may indicate fatigue failure.

Excessive sagging, twisting, or canting of stringers and floor beams must also be recorded. Observation of the floor system while traffic passes over the bridge assists in the inspection of these items. Where such conditions exist the connections along the members, particularly at the end points, should be observed for the effect upon the members to which they are attached. For example, excessive sag (or deflection) of a stringer may produce a transverse bending and subsequent lateral buckling of the floor beam to which it is connected.

Examples of floor systems in various conditions are shown in Fig. 6.4. A floor system in good condition is shown in Fig. 6.4a. No signs of damage or deterioration are evident in this system. Some minor paint and rust staining is beginning to show but the safe-carrying capacity has not yet been reduced. A floor system in fair condition is shown in Fig. 6.4b. The paint is no longer effective and rusting has progressed to the extent of possibly reducing the load-carrying capacity slightly. Fig. 6.4c shows a floor system in critical condition. The floor beam is sagging significantly and the deterioration of the entire floor system is in the advanced stages. Replacement or extensive rehabilitation is needed immediately.

C. Beams and Girders

1. Timber

Timber beams normally encountered during inspection are rectangular in cross section. They are usually placed on edge over their supports. When loaded, a downward deflection is produced placing the lower fibers in tension and the upper fibers in compression. This stressing is the opposite at any intermediate support when the beam is used over a continuous span. The beam area directly over the support points is subjected to vertical shear and crushing. Depending upon the unsupported length of the beam, the beam may be subjected to buckling or overturn if not laterally braced or blocked.

The stresses just discussed will reach their maximum values at definite locations in the beam:

At the center of the span the beam is usually at its maximum deflection. Horizontal shear may show along the longitudinal axis at the ends. Since the center point is receiving the maximum in bending, a transverse failure may occur across the upper or lower portion of the beam due to a compression or tension failure.

The bearing areas at the support points may show evidence of crushing or horizontal shear near the bearing point.

Either lateral buckling or overturn may occur anywhere along the beam. It may be most evident at the center of the unsupported length since this area is normally subject to the greatest deflection. Sighting along the length of the beam will usually disclose this lateral deflection.

Evidence of deterioration, such as decay, anywhere in the beam should be checked to determine its extent. Decay, while it may not visibly reduce

Fig. 6.4a Steel Floor System in Good Condition

Fig. 6.4b Rusting of Floor System Enough to Rate the Condition as Fair

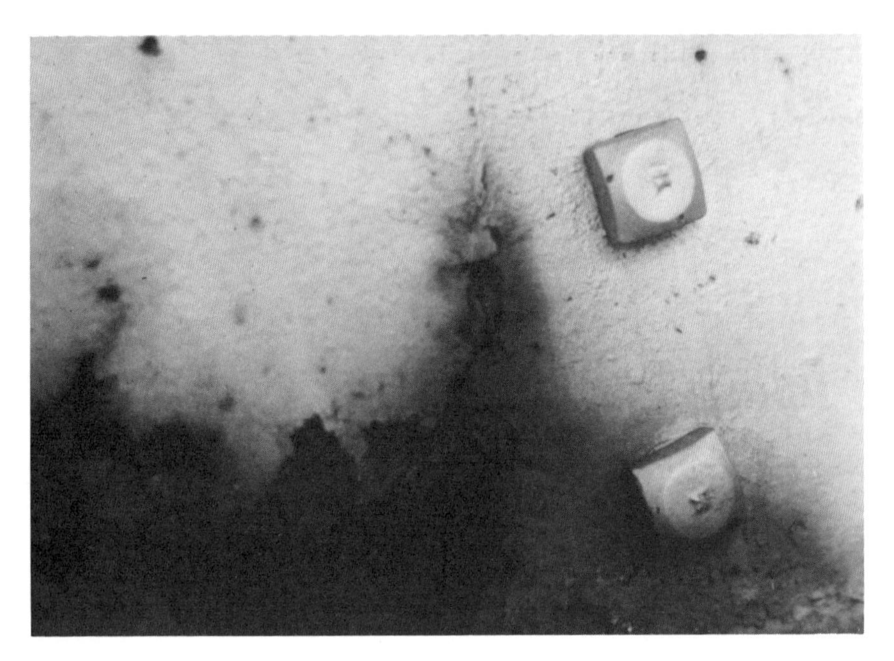

Fig. 6.4c Extensive Deterioration Merits Critical Condition Rating

the beam cross section, will reduce the beam's working ability to resist the various stresses produced under loading. Cross section measurements should be taken to provide the supervisor the dimensions of the available working cross section.

Crushing may appear at bearing points, either at the beam support points or where other bridge members, such as the stringer or decking, bear on the beam. The bearing point should be checked for the size of the bearing area and the extent of crushing.

Shearing failure along the grain of the beam or failure in the tension or compression fibers of the beam reduces the ability of the beam to perform its design function. The extent and location of the cracks or splits should be reported to indicate the remaining beam capability. Such shearing generally appears at the beam ends.

The veritcal and transverse alignment of the beam should be checked for abnormal deflection, particularly when live loads are on the bridge. Deflection usually indicates the start of beam failure in resisting normal design loads. It may also indicate failure elsewhere in the bridge, which shifts or transfers more of the load on the bridge to the sagging beams, i. e., overloading due to failure elsewhere. The amount and location of the sag should be reported.

The vertical and longitudinal alignment of the beam should be checked for overturning, buckling, or canting. As discussed previously in mechanics, Chapter 4, canting indicates that the upper portion of beam, being in compression, has a tendency to buckle. The effect of this lateral movement produces a rotational movement in the beam causing the canting.

The bolts, nuts, washers, or other devices such as split rings normally used in connecting timber beams should be inspected for looseness or deterioration due to corrosion or rust. The timber area immediately adjacent to the connector should be checked for crushing or possible decay, which may indicate later loosening of the connection. Debris at or around a connection should be noted and removed if possible to detect its effect on retaining moisture and contributing to decay at the connection.

Timber beams are relatively soft and susceptible to vehicle damage and other types of deterioration, which basically cause fiber damage and reduce the effective cross section. Likewise, timber is combustible and susceptible to fire damage, which also reduces the effective cross section and fiber strength.

Timber is generally more flexible than other bridge materials, such as concrete and metal. Inspection as the beam undergoes live loading is desirable in order to detect excessive flexing or movement and vibration. Evidence of abnormal displacement should also be noted in that the designed load distribution can be affected or may indicate a defect elsewhere in the bridge.

The inspection items discussed occur at different frequency during the life of a timber beam (or bridge). Since the durability of timber is relatively short in comparison with other bridge materials, deterioration is the most common form of timber beam defect.

Fig. 6.5 shows timber beams at various levels of the inspection rating. Timber beams in good condition are shown in Fig. 6.5a. No signs of damage or deterioration, canting or warping, or other problems are evident in this picture. The beams shown in Fig. 6.5b are in poor condition. Staining from water is extensive; severe damage and deterioration, particularly along the top of the beams, are evident. The beam shown in Fig. 6.5c is critical. The deterioration has extended to the bottom of the beam and is extensive enough to render the beam almost useless in transmitting shear forces to the pier cap. Immediate action is needed in this situation.

2. Concrete

The three basic cross-sectional types that will be discussed here are: the T type, the I type, and the Box.

Concrete slab types are discussed under the topic of decks, p. 83. Each of these types of beam or girder may be either reinforced or prestressed. Reinforcement increases the beam's capability to resist tension. Prestressing on the other hand adds compression stresses to the beam so that

Fig. 6.5a (left) Timber Beams in Good Condition; 6.5b (right) Deterioration of Timber Beam Merit Poor Condition Rating

Fig. 6.5c Extensive Deterioration of Timber—Critical Condition

the area in the beam subjected to tension will not exceed the tensional failure limit in that area when the beam or girder is loaded.

Beams or girders are normally braced by diaphragms or end dams as separate members to provide stiffening. In some cases the inspector may find that the diaphragms or end dams have been formed as part of the beam or girder when they were cast in place.

The most common stresses that occur in concrete beams and girders are essentially the same as those covered in timber beams. Loading of the simply supported beam or girder will cause bending and produce a compression stress in one flange and a corresponding tension in the other flange. The magnitude of the bending stress reaches its maximum at the center of the span and over the intermediate support of continuous spans. Maximum shearing stresses are produced at the supports or points of bearing upon the substructure. Shiplapped joints or bearing points on cantilevered girders must be examined as a unit for evidence of both the bearing shear at the support point and possible bending or compression stresses.

The significant areas in concrete beams and girders are essentially the same as those discussed for timber beams, namely, the center of the span, the bearing points, and flanges. However, the physical characteristics and properties of concrete plus the difference in the ultimate strengths of concrete compared to those of timber, present differences in both the appearance and the extent of stressed areas to be inspected. For example, small vertical cracks along the lower flange of a reinforced concrete beam or girder due to vertical deflection usually will be many in number and spaced over some length. These cracks are not considered in design and therefore are not significant in reinforced concrete. Conversely, a 45° crack near the end of a beam can be serious.

Concrete beams and girders contain either steel reinforcing rods or prestressing tendons. Surface cracks or spalling permit moisture to enter, and rusting may result along the tendons or rods. These longitudinal areas can become critical and therefore should be inspected carefully.

In Chapter 3, a section on the physical and mechanical properties of concrete, numerous types of deterioration were covered. However, the three most important deterioration conditions to look for during inspection of concrete beams and girders are scaling, spalling, and cracking.

Evidence of scaling usually leads to a slight reduction in cross section, and while not critical in itself, it may be significant if it appears near a support or bearing point.

Spalling, to a greater degree than scaling, causes reduction in cross section, particularly at the bearing points. Where spalling exposes reinforcing rods or prestressing tendons, the strength of the beam or girder may be significantly reduced. The severity of this condition applies along the entire length of the beam, and most particularly at the center of maximum flexure point, as well as at the end bearing point.

The entire beam or girder should be inspected for cracks. It may result from a fracture or indicate the possibility of a fracture and subsequent failure. The direction and extent of cracks or cracking should be recorded and described completely.

Alignment, movement, damage, and deterioration as inspection items are the same for concrete as for timber beams. However, in the case of prestressed concrete beams, a camber is produced by the prestressing tendons. Permanent camber may increase over time.

The pictures in Fig. 6.6 show several examples of concrete beams. Fig. 6.6a is a precast concrete beam in good condition. No cracking, spalling, or deterioration are evident from a visual inspection of the beam. The beam in Fig. 6.6b is in fair condition. Several horizontal cracks are visible, as well as stains indicating that water may be getting to the reinforcement. The cracks or deterioration are not severe enough to lower the safe load capacity, and hence the fair rating. The deterioration, cracks, and spalling are significant in the Fig. 6.6c, resulting in a poor rating. The reinforcement has been exposed to water and air, which will promote rapid rusting. The cracking and reinforcement deterioration would reduce the capacity rating, and call for a poor condition rating. The concrete and reinforcement deterioration is unusually severe in Fig. 6.6d. The condition of this beam is rated critical and requires immediate attention in terms of replacement or rehabilitation.

Fig. 6.6a Precast Concrete Beam in Good Condition

Fig. 6.6b Concrete Beam Rated Fair Because of Minor Horizontal Cracking

Fig. 6.6c Cracking and Spalling Reduces Condition Rating to Poor

Fig. 6.6d Severe Cracking, Spalling, Deteriorating of Concrete Give Beam a Critical Rating

3. Steel

Steel beams and girders react to loads in much the same manner as do timber and concrete beams and girders. The greater strength of steel permits use of less massive structural components for the beams and girders; however, the available working cross section becomes more important as an inspection item. The areas in which connections are made are also significant because of their effect on the available working cross section of the beam or girder.

Stresses in steel beams and girders are the same type as those encountered in timber and concrete beams. Over simple spans, the upper flange area of a steel beam or girder is normally in compression while the lower flange is stressed in tension. The opposite is true in continuous spans over intermediate supports. The web is designed to resist shear and is also subject to lateral and vertical buckling.

The significant areas in steel beams and girders are essentially the same as those in timber beams. Maximum bending is near the center of the simple span, producing compression and tension in the upper and lower flange areas, respectively. Maximum shear stresses will occur near the support points. These areas should be inspected accordingly.

The points at which connections are made are also important. Rivets or bolts require that holes be bored through the beam members. The effective cross section is reduced, and the load-carrying capability of the beam is reduced proportionately. In the case of welded connections, the quality and length of the weld are critical. Regardless of the method used in making connections, the connection must properly develop the required strength of the joint. Each joint is the result of design that is intended to maintain the structural integrity of the bridge structure. Inspection should, therefore, include close scrutiny of all connections to determine whether they are completely tight and capable of performing the function for which they were designed. Composite beams have not been discussed as a separate type of beam or girder. The concrete portions are inspected in the same manner as concrete beams, and the steel portions, the same as steel beams or girders are inspected.

Steel beams and girders should be inspected in the same manner as other steel bridge members. During inspection, the inspector must consider the types and locations of the stresses developed when the beam or girder is loaded and the ability of the beam or girder to resist anticipated loading.

The extent of metal deterioration due to corrosion or rust should be investigated to determine its effect on the beam or girder cross section. In severe cases of corrosion, particularly along the web and flange interface, measurements should be taken and recorded as part of the inspection report in order to assist the bridge engineer in their initial review and in further analysis, if required. Built-up beams or girders should be inspected along their joining surfaces for evidence of rust and deterioration.

Visual observation of the beam or girder will usually disclose any distortions in the beam or girder components. Sighting along the length of the member will also disclose sagging, buckling, or other irregularities that must be recorded. Connections in the vicinity of the distortion should be inspected for overstress or eccentricity in load transmission and possible future failure. The cause for the distortion should be recorded if possible, as well as the effect produced in members affected by the distortion.

Riveted, bolted, and welded connections should be inspected for completeness, looseness, signs of movement, and evidence of cracking or buckling in the vicinity of the connection. Where sudden changes in cross section occur, such as flange-cover plate ends, the connection should be inspected to ensure that the components are fully developed. Dirt and debris have a tendency to collect in the vicinity of connections. Such debris should be removed and the connection inspected for deterioration due to rust or corrosion. In those instances where the debris cannot be removed, this fact should be reported.

Collision damage to beams and girders is usually apparent. Impact points should be inspected for evidence of cracks, breaks, or tears in the beam, damage to connections, and the extent of distortion in the member itself, as well as other portions of the bridge. Fire damage will usually result in the development of pronounced sagging of the members that have been subjected to extreme heat. Frozen joints, hangers, or obstructions to thermal expansion and contraction will produce damage either at that point or cause overstress elsewhere in the bridge member.

4. Bracing

A bridge usually consists of two or more parallel main carrying members supporting the bridge floor or roadway. These members may be girders, trusses, or beams. In the case of girders or trusses, the bridge may be either a deck or a thru type. These two members are tied together by bracing. Bracing in a horizontal plane between the two lower chords or flanges is termed the lower lateral bracing system, and similarly, the bracing between the upper chords or flanges is known as the upper lateral bracing system. Bracing in a transverse vertical plane between opposite main carrying members is called the sway bracing for truss bridges or cross frames for girder bridges, while the bracing in the plane of the end posts on through trusses is termed the portal bracing.

The DOT Manual [1] has divided the types of bracing into three general categories: diaphragms, lateral bracing, portals, and sway frames.

Several examples of bracing are shown in Fig. 6.7. Fig. 6.7a shows sway bracing for a truss bridge. The bracing is in good condition with no evidence of damage or deterioration. Fig. 6.7b shows lateral bracing and sway bracing for a truss bridge. The lateral bracing forms an "X" between two adjacent panels of the truss. The sway bracing is the K brace laying on its back in this figure. Again, the bracing is in good condition.

Fig. 6.7a Typical Sway Bracing for a Thru Truss Bridge

Fig. 6.7b Typical Lateral and Sway Bracing for Deck Truss

Diaphragms are transverse members between beams, girders, or stringers, which brace or stiffen these main members. Diaphragms also help distribute the load between adjacent beams and where used at bridge deck ends, diaphragms also may function as end supports for the bridge deck. On deep girder or truss bridges, cross frames serve the same purposes. On concrete bridges they may be cast with the beams; or, if precast concrete beams are employed, they are cast in place transversely between the beams and tied through with cables or rods, or cast with the deck.

Lateral bracing consists of lateral diagonal members and perpendicularly transverse members termed lateral struts. This strut also serves as part of the cross frame. When the chords or flanges of the main supporting members are horizontal, the lateral bracing is usually a horizontal truss itself, the chords of which are the chords or flanges of the main members.

Transverse bracing at the panel points in truss bridges in a vertical plane is termed sway bracing. Sway bracing can be accomplished by X, K or knee-bracing members connecting the lateral struts to the bridge panels or verticals. Sway braces may use the lateral strut as one of the chords in a transverse truss. In this instance the sway bracing is usually called a sway frame. Portal bracing is a special type of sway bracing employed with thru-type bridges in the plane of the portal itself. Portal bracing is generally larger or heavier because of the magnitude of the loads or stresses at the hip joint and in the end posts. The important aspect of portal bracing for inspection purposes is the fact that the shape and dimensions of the portal determine the size vehicle which the bridge will accommodate. Similarly, the portal dimensions normally dictate or limit the sway bracing dimensions at the intermediate panel points. Accordingly, portals and the lower portions of sway bracing are subject to vehicle collision damage, a significant item which will be covered in more detail later.

a. Lateral Forces The effect of the wind blowing on or against a bridge produces a wind load on the bridge. Wind blowing on vehicles crossing a bridge also produces a wind load on the bridge. For practical purposes, to simplify bridge design, this loading is considered to act horizontally with transverse and longitudinal components. For inspection purposes, it also facilitates visualizing the functioning of bracing as the bracing reacts to counter this force.

The weight or deal load of that part of the superstructure transmitted to the bridge panels, beams, girders, or trusses causes deflection of transverse floor beams; tending to rotate the bridge panels in opposite directions. Live loads further deflect the floor beams and in turn tend to rotate the bridge panels in the same manner as do the dead loads. In addition, these loads tend to cause vibrations in the bridge as they pass along the bridge over successive deck support points or floor beams.

b. Stresses Stresses in bracing members occur in much the same way that they occur in a truss panel in which a diagonal and a counter are found. In trusses, a counter is placed in tension in a panel with a diagonal to ensure there is no stress reversal in the diagonal. As the chord or flange portions of the bridge panels are placed in tension or compression, the lateral braces prevent longitudinal movement of the panels and preserve the rectangular configuration within each panel section. For practical purposes, bracing inspection can be considered similar to inspection of truss members in that here also the tension and compression stresses will occur along the longitudinal axis of the brace. Diaphragms are inspected for their stiffening effect on the main carrying members, although girder bridges may use knee braces with solid webs connected to the stiffeners or main members and the floor beams to achieve the desired bracing.

Design practice has usually leaned toward heavier bracing than is required when stresses were computed during the design. This practice has resulted in generally increased stiffness in the bridge and reduced maintenance costs, as well as simplified fabrication procedures since the variety of bracing sizes is reduced. For inspection purposes, this tendency fairly well eliminates concern for the utlimate strength of the bracing member unless alignment, deterioration, or connection conditions indicate otherwise.

c. Materials Bracing members are usually made of the same material as the bridge itself, i.e., steel girder or steel trusses normally use steel bracing members. However, many timber bridges use steel rods or bars for bracing. These are primarily used as diagonals or as lateral bracing. In the case of cast-in-place concrete bridges, the braces are cast with other portions of the span such as stringers, beams, or slabs. For prestressed concrete beams, as mentioned previously, cast-in-place concrete diaphragms are tied between the beams by means of cables or rods running through the diaphragm.

5. Connections

Bracing on steel beam, girder, and truss bridges is normally connected to the main supporting members. The connection method may be directly to a main member or, as in the majority of cases, by means of plates or angles. Fastening devices may be bolts, welds, or rivets. Where the bracing member is a rod or eyebar, the loop ends are generally pin or bolt connected or the rod is threaded and bolted. In the case of lateral bracing, the connection generally includes connection through a gusset plate to both the main member and the lateral strut or transverse member, such as a floor beam.

Timber braces are generally either bolted or spiked directly to members or both. Struts which bear on other members may be provided with metal end bearings.

In the case of concrete slab bridges, the rigidity or the deck slab may eliminate the need for separate and distinct bracing. In the case of cast-in-place concrete beam bridges, the diaphragm is usually cast monolithically with the beams or slab. Reinforcing rods provide the necessary connection. Cast-in-place concrete diaphragms have already been discussed.

a. Configurations Single angles are often used for the bottom laterals in lighter bridges and double angles or structural Ts in heavier bridges. The longer leg of the single angle is used for connection. Where double angles are used, the longer legs are normally joined and the bracing is connected at the shorter legs of the brace. The laterals of the top chord bracing for heavier trusses may be made of rolled wide-flange sections or built-up members, usually the same depth of the chord member itself. Transverse sway bracing may have either two or four angles at the top and two angles at the bottom with a system of diagonal angles or built-up members between.

The bracing members form a transverse vertical or horizontal plane between the main carrying members of the bridge. When viewed separately, they can be considered a type of truss. Sway bracing, however, frequently entails knee braces. A knee brace provides stiffness at the brace point in addition to acting as a sway brace. The rigidity provided by knee bracing in transmitting loads makes inspection of the members to which the knee brace is attached important. Bending may occur in these members at or near the brace point. Triangular plate stiffeners are frequently used in plate girder construction and function in the same manner as knee braces. For the most part, this type of brace would be found on thru girder bridges.

b. Significant Areas The significant areas in bracing members and systems are similar to those in trusses and, to some extent, girders and beams. Since bracing members act in tension or compression, and in some cases, bending, the cross section of the member in both area and shape is significant. Reduction in area, regardless of cause, obviously reduced the capability of the member to resist the force or load it was designed to carry. Similarly, a change in the shape of the cross section of the member also reduced the member's load-carrying capability.

The point at which the bracing is connected to the bridge main members and other bridge members is significant. Connections are designed to transmit loads along the axis of the bracing member and, also, to or from the bridge members themselves. Deviation from this alignment along the axis of the brace may produce bending in the brace and may reduce its capability. Misalignment at the connection may also induce an overstress in the connection plate or shear in the rivets or bolts used in the connection as well. Looseness at the connection will produce damage due to vibration, impact, and the associated abrasion of the parts wearing against each other. The bracing members will also be subject to additional fatigue stress due to the vibrations and excessive movement at the loose joint connections.

Overhead bracing members may suffer considerable damage from vehicular collision. The clearance restriction areas can, therefore, be classed as critical for inspection purposes. Fortunately, most vehicle or collision damage is readily apparent and easily noted, but the effect of the damage may not be easily determined. When such damage has taken place, new and additional stresses are placed on the bridge members to which the bracing member is or may have been attached. These effects can usually be determined by noting the amount and type of movement produced in these other members by the disfigured brace. Additionally, when bracing members are broken, they cannot receive or distribute loads properly with respect to the member to which they were connected.

Wind shoes and lateral shear keys may be used to distribute wind loads from one member to the next adjacent member or from the main lateral bracing system to the supporting substructure and should be checked.

c. Deterioration and Damage Braces and bracing members are susceptible to the same types of deterioration that other bridge members are subject to. Depending upon the type of materials used, braces are subject to corrosion, rust, cracking, spalling, splitting, and decay. Inspection must include the proper identification and location of deterioration plus adequate recording and description in the inspection report. This description along with the inspector's evaluation will provide supervisory personnel with the information they need to evaluate the bridge as a whole.

Damage to bracing not only reduces its capability to perform its primary function but also affects other bridge members or elements. Inspection reports should, if possible, include the probable cause of damage, to better enable the bridge engineers to assess the effect of the type damage on the bracing member and related bridge members. For example, fire damage on prestressed concrete beams may have a different effect than on reinforced concrete beams. These effects are related to the number and location of the rebars or prestressing tendons and the difference between the physical properties of each type of steel. Possible secondary effects would be cracked decks due to sag in the span, overturn of rocker bearings, or misalignment such as warping or torsional bending.

6. Bearings and Bearing Plates

The principal purpose of bearings and bearing plates is to transmit the weight of the bridge and the loads it supports to the bridge supports or abutments. Since the magnitude of such loads is large, the bearings or bearing plates are normally constructed of high-strength material, usually metal.

The secondary purpose of bearings and bearing plates is to permit rotation of the bridge relative to the support or abutment, while transmitting the load to the abutment. In addition to permitting rotation, certain types of bearings must also permit longitudinal movement of the bridge in relation to the support. These movements will be discussed later as each type of bearing or bearing plate is considered.

a. Functional Components Sole plate is generally a rectangular piece of material large enough in area to support the bridge member at the bearing area without adversely affecting that portion of the bridge main supporting member (truss, girder, or beam), at the bearing area. It is attached to the main supporting member.

Masonary plate (bearing plate) is similar to the sole plate except that its area is large enough to preclude damage of the support or bridge seat at that bearing area. It rests upon the bridge seat.

Bearing (or bearing surface) is that portion of the bearing between the sole plate and masonry plate which permits rotation and longitudinal movement of certain types of bridge structures at that bearing point.

Figure 6.8 shows the three parts: the sole plate, the masonry plate, and bearing. This type of bearing is common. It acts as a hinge; the beam end is allowed to rotate but no horizontal or vertical movements are permitted.

Fig. 6.8 Bridge Support Showing the Sole Plate, Masonary Plate, and Bearing

Not all bearings will have the three separate and distinct components just discussed; however, at least one, the bearing surface, is always present for inspection. Some bearings may have more than three parts.

b. Movement at Bearings As loads are applied to the bridge, such as moving vehicular traffic, the bridge span is flexed or deflected downward. A resultant rotation occurs at each end point or support. This rotation must be accommodated at the bearing, or additional stress is produced in the end of the span member at the bearing point. This applies normally to spans greater than 50 ft.

Two types of hinge bearings are shown in Fig. 6.9. These bearings allow only rotation of the end of the beam. Fig. 6.9a is a pedestal- and shoe-type bearing. Fig. 6.9b shows a pad or plate-type hinge bearing.

The principal cause of longitudinal movement at bearings is temperature changes that produce expansion or contraction of the bridge span in total overall length. As the expansion or contraction occurs, the bearing must accommodate this movement with as little resistance or friction as possible in order to prevent additional stress in the end of the span member or on the abutment.

Fig. 6.9a Hinge Bearing—Pedestal and Shoe Type

Fig. 6.9b Hinge Bearing—Pad or Plate Type

Three types of bearing connections that allow horizontal movement are shown in Fig. 6.10. Fig. 6.10a is a rocker bearing, which is pinned at the top and moves back and forth on the curved base. Fig. 6.10b shows a roller bearing. A hanger-type connection is shown in Fig. 6.10c. This connection is away from the piers to take advantage of better stress distribution and drainage conditions from the deck and still allow horizontal movement.

Transverse or lateral movement is prevented by keys or locks, thus preserving the longitudinal alignment of the bridge over the bearings and the floor system or roadway.

c. Types of Bearings There are two general categories of bearings: fixed and expansion. Fixed bearings do not permit horizontal movement of the bridge or span at the bearing point. Fixed bearings do, however, permit rotation at the bearing point due to deflection of the span. Expansion bearings permit both rotation and longitudinal movement at the bearing point. Each of these features will be discussed for the particular type of connection.

Sole plates are generally permanently attached to the bottom flange of girders and beams or the bottom chords of trusses. Attachment may be by rivets, bolts, or they may be welded to the lower portion of the bridge span

Fig. 6.10a Expansion Bearing of the Rocker Configuration

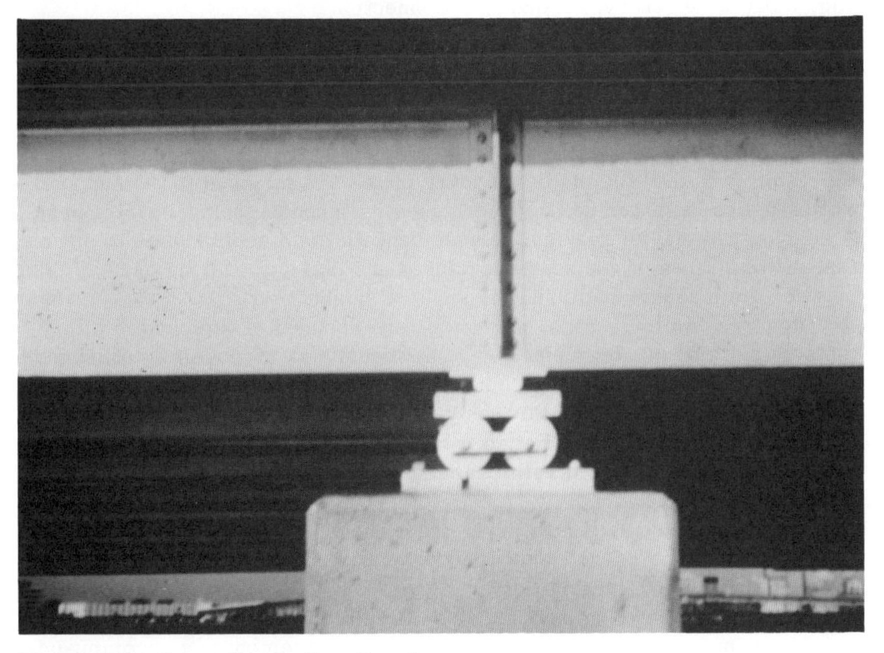

Fig. 6.10b Expansion Roller Bearing

Fig. 6.10c Hanger Type Expansion Connection

end. Where concrete beams, girders, or slabs are used at the span end, the lower flange or portion of the beam, girder, or slab may function as a sole plate.

Similar to the sole plate, masonry plates are attached to the supports or abutments by either bolts or permanently imbedded at the bearing seat. A bearing pad may or may not be used beneath the masonry plate to ensure that the total area of contact is developed and that a good bond or seal exists between the masonry plate and the support or abutment.

Rockers, rollers, pads, and plates are secured or attached to either the sole plate or masonry plate by various methods to permit rotation and/or longitudinal movement at either the upper or lower part of each (or both). For example, slotted holes through which anchor bolts pass, permitting horizontal movement, are limited by the length of slotted holes in rockers or bearing plates or keyed rollers that roll positively as the slotted sole plate moves.

Where pins are used at bearing connections, rotation must be provided at either the pin or at the end of the pin support. This element could be a combination pin and rocker bearing.

Elastomeric bearing pads provide a flexible bearing surface at the span bearing points. Both rotation and longitudinal movement are accommodated at the bearing point by the elastic deformation of the pad. Pad size, particularly thickness, is determined and designed to provide for such expected movements at the bearing point.

d. Friction With the exception of elastomeric pads, the friction produced at the bearing points is kept to a minimum. In the case of flat or plate surfaces, the bearing areas are planned to provide as smooth a surface as is possible with the materials used for longitudinal movements. In the case of friction at rotational bearing points, rounded portions are machined as true to the expected rotational radius as possible.

Lubricants such as graphite are used on bearings where two parts move against one another to further reduce friction and subsequent wear, just as one would oil a door hinge or a wheel bearing. Solid lubricants are also now being employed to great advantage, in that maintenance and continued lubrication are eliminated, and hence, there is greater economy. Notable in this area is teflon (TFE), which tolerates extremes in temperature and pressure while maintaining its low frictional resistance to movement. Current developments have extended to use of coatings on curved surfaces, both cylindrical and spherical, as well as on flat or planar surfaces. Flat plates are usually made of lubricated bronze or copper.

e. Inspection Items The various types of deterioration in bearings can be observed and determined by applying that knowledge already gained about deterioration and the mechanical properties of materials. For example, steel bearing plates are subject to the same corrosion-rusting action that steel girders or beams are subject to. In other words, deterioration is based on the material of the bearing and the adverse effects of such things as moisture or corrosive acids. In most instances, deterioration is visible in the form of rust, corrosion, or evidence of "frozen" joints or connections, which are evidenced in the extreme by bending, buckling, or improper alignment of members. Elastomeric material deterioration generally develops as cracks, splits, or tears due to overload or excessive stress.

Associated with deterioration by rusting or corrosion is accumulation of dirt and debris at the bearing point, which is the prime contributor to and secondary cause of this condition. Such debris has a tendency to absorb and hold moisture of corrosive material in close proximity to the bearing surfaces. This proximity hastens the corrosive action. In the case of rockers, or slotted plates, debris may become wedged or impacted at the bearing surfaces and movement of the bearing itself is restricted. An example of rusting and debris at a bearing is shown in Fig. 6.11.

Each bearing is designed to perform a specific function at the bearing point. For the various types of bearings, each has a characteristic size, shape, and configuration corresponding to its specific function. Movement of the bridge or span at the bearing point will either be rotational or longitudinal or both. Proper alignment of the component parts of the bearing

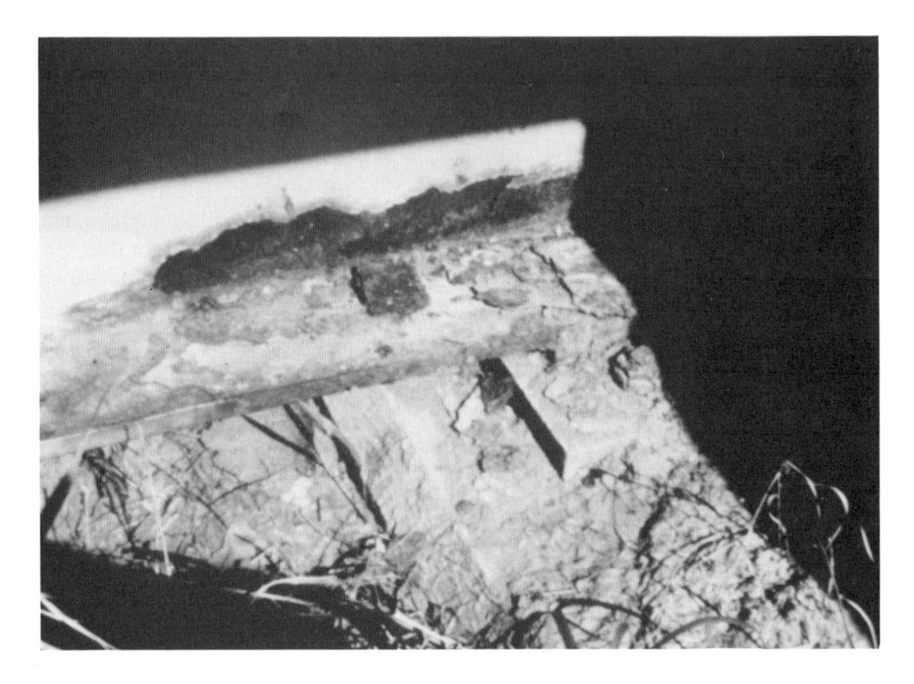

Fig. 6.11 Debris Collection Around Bearing Support

permits this movement with a minimum of resistance. Where bearings are not properly aligned, the normal movement at the bearing points introduces new or different stresses in the bridge span at the support points, which, in turn, will be transmitted to other members of the bridge.

Similarly, movement of the bridge span over a period of time may tend to cause parts of bearings to creep or move out of place. While transverse movement is normally restricted, constant vibration and repeated live loadings may cause movement of bearing pads or bearing plates. In many instances, reference points or marks are scribed, punched, or cut into adjacent parts of moving bearings in order to provide a gauge of movement that has occurred at the bearing point by either rotation or horizontal travel of the parts in relation to each other.

The size and shape of bearings and bearing components are determined, among other factors, by the allowable strength of the materials used per unit of contact area at the bearing point. Inspection of the bearing must include a determination that the full bearing surface is in contact. Where only partial contact is disclosed, damage can occur to the span, the support, the bearing, or all three. For example, where a girder span may have moved horizontally and the sole plate rests on only a portion of the masonry plate, the load transmitted to the abutment may exceed the bearing value of the masonry used in the abutment and result in crushing of the support. Conversely, buckling in the web of the girder could also result if the load being supported by only a portion of the bearing surface results in a bearing pressure that exceeds the allowable value for the steel used in the girder.

Loose bearings may be detected by noise at the bearing when the bridge is subjected to live loads such as traffic or movement of the bearing parts. Among the various causes of looseness the inspector should look for are:

Settling or movement of the bearing support away from the portion of the bridge supported.

Excessive rusting or corrosion, which results in loss of material in the bearing itself.

Excessive deflection or vibration in the bridge span members themselves, such as broken or lost members.

Loose fasteners such as rivets or bolts used to attach the bearing to either the bridge span or the bridge support.

In the case of bolts, nuts may be missing completely, or not properly tightened or adjusted. Worn bearing members such as pins or pin shoes may be loose or missing.

Displacement of elastomeric bearings or enough deterioration to affect the bearing function will normally be discovered by the inspector before looseness would be evidenced. Such deterioration might be tears or excessive and uneven bulging. Also, the large compressive forces exerted on elastomeric pads generally preclude looseness alone as an important inspection item for these bearings. Fig. 6.12 shows deformation and bulging of an elastomeric pad.

Hangers are used on certain configurations of cantilever spans. Acting similar to chain links, they permit both rotation and longitudinal movement at the span connection. Hangers develop stresses in tension and should be inspected accordingly.

The pins and pin connections are inspected for wear, looseness, and proper attachment. Hangers basically consist of two plates. The area between is difficult to inspect and is vulnerable to deterioration and stress corrosion of the pin connections, as well as accumulation of dirt and debris.

Windlocks and keys, while not bearings in themselves, prevent or restrict lateral or transverse movement of the span. In so doing, they assist in maintaining the alignment of the span over the bearing points while permitting rotation and longitudinal movement at the bearings. Lateral shear keys are particularly important where the bridge is skewed or located at a curve.

Fig. 6.12 Elastomeric Pad Type Bearing

Fig. 6.13 shows a series of bearings in variously rated conditions. Fig. 6.13a shows a rocker bearing in good condition. The rocker is free to move, is well aligned, free of debris, and no signs of rusting are evident. Fig. 6.13b is a rocker in fair condition. The alignment is suspect, and stains are visible indicating that rusting has begun. The rocker in Fig. 6.13c is in poor condition. The beam is resting against the abutment face and the beam is no longer free to move. Fig. 6.13d shows a roller bearing in critical condition. The bearing is not stable, and no movement is possible without causing failure at this location.

7. Railings, Sidewalks, and Curbs

Railings, sidewalks, and curbs normally do not contribute to the structural strength of the bridge and are provided mainly for public safety. Consequently, the design of railings from the standpoint of their capacity to withstand or absorb vehicle impact is not of greatest importance here. The inspector's job will be to inspect and report the present condition of these elements with regard to their ability to meet the safety standards for which they were designed or current standards. Design is for others. Inspectors should only consider these elements in regard to the original design or new standards provided by the design personnel. The primary concern is the

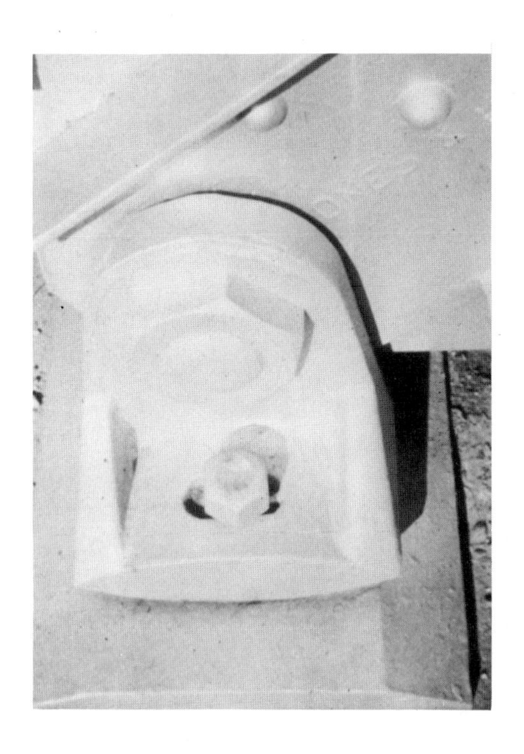

Fig. 6.13a Rocker Bearing in Good Condition

Fig. 6.13b Rocker in Fair Condition Because of Alignment, Rusting, and Water Stains

Fig. 6.13c Rocker in Poor Condition—No Expansion Movement Available

Fig. 6.13d Unstable Roller Bearing is Rated in Critical Condition

likelihood of the elements to perform as intended in view of deterioration, distress, or collision damage.

a. Railings A guard rail is a fencelike barrier, or protection, built within the roadway shoulder area and intended to function in combination as a guide and protective restraining guard for the movement of vehicles or pedestrian traffic and to prevent or hinder the accidental passage of such traffic beyond the berm line of the roadway. The berm line is defined as the location where the top surface of an approach embankment or causeway is intersected by the surface of the side slope. A concrete bridge railing is shown in Fig. 6.14.

Railing is also wooden, brick, stone, concrete, or a metal fencelike construction; it is built at the outermost edge of the roadway or sidewalk portion of a bridge to guard or guide the movement of both pedestrian and vehicular traffic and to prevent the accidental passage of traffic over the side of the structure.

The term handrail is commonly applied only to railing presenting a latticed, barred, blaustered, or otherwise open web construction.

Fig. 6.14 Concrete Bridge Railing

By definition, there is no major difference between guard rail and railing. Both function to protect and guide the movement of traffic. Any difference occurs as a result of the location of the railing. Approach guard rail is a specific application of the term guard rail indicating that it is on the bridge approach. Bridge railing is a specific application of the term railing.

As pointed out previously, it is not appropriate for the inspector to report upon the adequacy of design unless he or she has been properly trained in making such observations by competent authority. The inspector will certainly report instances of railing penetration or damage caused by collision, but an opinion on the adequacy of design is best left to bridge and highway safety engineers. Report the results of the inspection of railings in terms of condition, that is, good, fair, poor, or critical. A bridge railing that has a section missing may present a critical condition, either as a snagging hazard or in regard to reducing the penetration hazard. If the bridge plans call for an approach guard rail and a section has been penetrated, this would be reported as a critical approach guard rail condition.

b. Sidewalks and Curbs Sidewalks and curbs are normally provided to protect pedestrians crossing the bridge. Specifically, sidewalks are those portions of the bridge floor area serving pedestrian traffic only. Curbs may be of stone, concrete, steel or timber. They parallel the side limit of the roadway to guide the movement of vehicle wheels and safeguard bridge trusses, railings, and other constructions existing outside the roadway limit, as well as protect pedestrian traffic upon sidewalks from collision with vehicles.

A safety curb is a narrow curb between 9- and 24-in. wide, serving only as a refuge for pedestrians on a bridge. Pedestrian traffic is not normal for bridges with only safety curbs. A wheel guard is a timber piece placed longitudinally along the side limit of the roadway to guide the movements of vehicle wheels and safeguard the bridge trusses, railings, and other constructions existing outside the roadway limit from the collision with vehicles.

When required, sidewalks may be placed upon cantilever type supports, which are commonly triangular in shape. These supports, known as sidewalk brackets, are generally attached to a girder, truss, or bent. Sidewalk stringers are placed on top of these brackets, with sidewalk and railing resting atop the stringers. Sidewalk brackets may be constructed of concrete, steel, or timber. As part of the inspection of sidewalks, these brackets should be checked for structural integrity.

Approach guard rail is normally inspected with the bridge approach. However, the inspector should consider that the approach guard rail is a subsystem, along with the bridge railing, of an integrated system designed to protect and guide vehicular and pedestrian traffic. Bridge railings may be constructed of concrete, metal, or timber. Most timber systems are being replaced with concrete or steel. Properly designed railing systems will reduce the hazards of penetration, snagging, bridge end collision, and collisions as a result of railing redirection of vehicles. Sidewalks and curbs may be constructed of concrete, metal, or timber, and are normally for the protection of pedestrians.

8. Expansion Joints

Expansion joints employed on bridges have many of the same characteristics as bearings in regard to movement. Expansion joints permit movement of the bridge without damaging the deck, abutments, or approaches yet provide continuity of the deck. Also, like bearings, there is a variety of types of bridge joints that can be encountered when making on-site inspections.

The discussion below considers various types of expansion joints, their configuration, types of materials used, and the functional aspects of these joints so that the significance of specific inspection items can be understood. Also, drainage as it pertains to expansion joints will be discussed.

Expansion joints are primarily used at the deck slab ends to permit the thermal expansion and contraction of the slab. On long bridges, additional expansion joints may be used at intermediate points. The relationship of temperature ranges to span length is considered during bridge design and the amount of expansion and contraction required to allow for temperature changes is calculated in order to ascertain the proper type of expansion joints. The expansion joint opening or joint filler material permits a calculated amount of movement before crushing or buckling occurs at the joint sides.

Expansion joints can be categorized in a number of ways. The DOT Manual [1] lists them as open, poured sealant and joint fillers, compression seals, steel expansion plates, and steel finger dams, each according to expansion distance.

The open joint may be found where the expansion distance is comparatively small. If water tightness is not required to prevent drainage from falling on substructure elements or bearings, the joint can be left open and normal expansion, and contraction can take place.

Joints may be filled with a joint filler or a sealant and joint filler. As the bridge expands or contracts, the material in the joint is compressed or stretched. Such joints should be watertight and the filler material is designed to perform this function. Debris may prevent joint movement if the filler fails, and may damage the joint sides or joint material. It may spall either or both sides of the jointed slabs, or cause overstress in other bridge elements, e.g., at the expansion joint near the abutment, main support member, or at the other end of the span at the abutment, bearing, or deck slab joint.

Most types of debris tend to retain moisture and hence contribute to the deterioration of adjacent bridge material. Such moisture, particularly that containing chemical deicing agents, hastens corrosion or rusting of steel, and in the case of concrete, scaling. When debris collects in drain troughs under open expansion joints, overflow drainage on abutments and bearings can cause deterioration at these points.

Deterioration of sealed expansion joints usually results in loss of joint material. Loss of joint material, in turn, reduces the capacity of the joint to absorb slab movement and maintain watertightness in the joint. Further deterioration produces disintegration of concrete and permits deck drainage water to fall on bearings and bridge seats. Corrosion of expansion joints such as sliding plates or finger joints may lead to "freezing" of the joint's moving parts, or conversely, further open the joint, making it more susceptible to collecting debris. Where anchorages or armoring is used in the joint, corrosion can cause looseness and further damage by spalling and disintegration of the concrete or pavement.

Excessive wear and fracture at expansion joints indicates deterioration of the joint material, usually from some type of abnormal functioning of the joint or external causes. Evidence of sealant separating from the joint edges or shriveling and cracking of the seal indicates the inability of the seal to expand or return to its original shape as the slab contracts and joint opens. Such openings permit foreign matter, often incompressible, to enter the joint and cause further damage to the joint. Overfilling of joints with sealant may produce a bulge when they are compressed by expansion of the slabs. Such bulging above the surface of the roadway makes the sealant vulnerable to traffic impact; consequently, increased wear or fracture may occur.

Bent, cracked, or broken fingers on finger dams may be the result of traffic damage, poor alignment, or loose anchorages and may indicate lateral movement of the deck slab. Subsequent movement of the joint can cause additional damage to the joint as well as produce localized stresses due to misalignment of the fingers.

Fractures and spalling of the pavement or deck in the area adjacent to the joint may cause subsequent failure of the joint by loosening joint side-support material. This looseness can contribute to further deterioration from the two causes previously discussed.

These inspection items have been discussed with regard to expansion or contraction by the bridge or deck. Movement of the abutment must also be considered when inspecting joints. Movement of the abutment may decrease or even completely close the joint opening, preventing free expansion of the bridge. Conversely, such movement may increase the joint opening.

In the inspection report the movement or opening distance and temperature must be recorded at the time of inspection, to provide supervisory personnel with adequate data upon which to base an overall evaluation of the bridge and to determine the urgency of additional investigation of existing conditions.

Fig. 6.15a Steel Finger Joint in Good Condition

Fig. 6.15b Sliding Plate Joint Rated Fair Because of Looseness and Deterioration

Fig. 6.15c Sliding Plate in Poor Condition Because of Deterioration and Debris

Fig. 6.15d Sealed Concrete Joint in Poor Condition—Seal is Loose and Spalling Evident

Several examples of bridge joints are shown in Fig. 6.15. Fig. 6.15a shows a steel finger joint in good condition. The joint is free to move and no damage or deterioration are evident. A sliding plate joint is shown in Fig. 6.15b. The joint is free to move but appears to have a gap near the railing, and some deterioration around the joint is visible. Another sliding plate joint is shown in Fig. 6.15c. This joint is in poor condition, with a collection of debris in the joint and some deterioration around the joint. Fig. 6.15d shows a sealed concrete joint. The seal is no longer effective, and some spalling of the joint is evident.

References

1. U.S. Department of Transportation, Federal Highway Administration, Bridge Inspector's Training Manual 70, Washington, D.C., 1971.
2. American Association of State Highway and Transportation Officials, Standard Specifications for Highway Bridges, 12th Ed., Washington, D.C., 1977.
3. American Association of State Highway and Transportation Officials, Manual for Maintenance Inspection of Bridges, Washington, D.C., 1978.
4. U.S. Department of Transportation, Federal Highway Administration, Recording and Coding Guide for the Structure Inventory and Appraisal of the Nation's Bridges, Washington, D.C., January, 1979.

Chapter 7

INSPECTION OF THE SUBSTRUCTURE

I. Soil-Foundation Interaction

Regardless of how the substructure or superstructure is designed, the total weight of the bridge plus the loads which it must be able to carry are supported by the underlying foundation soils. In reviewing the bridge plans, the inspector must focus on the foundation support material upon which the bridge rests. He should be informed about the nature and physical properties of the soils that hold and support the bridge, enabling it to withstand the forces of nature (flood, winds) as well as the forces of human invention (vehicle loading, construction projects).

For some time, engineers have known that foundation materials vary considerably in their composition and their inherent properties. All too often, however, little attention has been devoted to the subject called soil mechanics. Instead, extensive safety factors have been incorporated into the original design with the hope that these measures would be insurance against the consequences of inadequate or inaccurate foundation data and ultimately, against actual bridge failure. This approach has proven unsatisfactory.

In more recent years, soil sampling, testing, and analysis techniques have improved greatly and the whole science of soil engineering has been taken more seriously.

The foundation, then, is really the "invisible" part of the bridge, for it lies buried beneath the surface. A large portion of the money spent for constructing bridges is for what lies below the ground. The design and allocation of foundation is, not surprisingly, a specialty requiring highly trained engineers and soil specialists.

From the inspection standpoint, the basic physical properties of these support materials, along with the causes of foundation movements, and how they affect the structural integrity of the bridge, are of primary importance.

A. Nature of Soils

There are three general classifications of soil the inspector needs to consider. They are glacial soils, alluvial soils, and residual soils.

Glacial soils are most common in the northern United States. They tend to be very irregular, with layers varying considerably. Sometimes poor soils that contain sand and soft clay are located under more stable ones containing hard deposits of stiff clay, gravel, and rock.

Alluvial soils are usually found in flood plains and the delta areas of river basins. This type of soil is readily found along the Mississippi River Valley and at the mouth of the river. Normally, alluvial soils contain large amounts of fine sand and silt and are deposited in thin layers that may vary considerably in composition.

Residual soils are those soils formed by the weathering of rock. They may be sands, silts, or clays, depending on the amount of weathering that has occurred and the type of rock from which they are derived. For example, limestones and shales often form clayey soils, while sandstones and granites often form sandy soils. Local rock formations can be studied to determine what types of residual soils they produce.

Compact, well-graded sand is normally a good foundation material, provided it contains no soft-clay layers. Loose, saturated, poorly graded sand has very little bearing capacity. Foundations placed on compact sand must be protected from scour. Failures of several structures have been attributed to the flow of unconfined, fine sandy soil. Piers for a bascule span at Bridgeport, Connecticut, rested on 25-ft piles driven into material classified from wash borings as fine sand. The load on each pile was 27 tons. The piers tipped in opposite directions, causing additional settlement of over 1 ft with some movement downstream. The failure was caused by piles being driven into fine, unconfined, submerged sand. This failure could have been prevented by obtaining soil samples to determine the true nature of the soil layers and then driving the piles into the gravel bed or to the rock below the sand, as was done for some nearby piers.

Clay usually contains water and will effectively retain water. When it is dried it becomes hard and brittle and shows signs of shrinking and cracking. Yet upon adding water to dry clay, the clay tends to become highly plastic and compressible. If pressure from a load is applied continuously over a long period of time, it may cause the clay to flow. Surface water will soften and erode clay easily.

The physical properties of clay vary greatly. The failure to appreciate this fact has been a source of major foundation trouble. Some clay particles can take on large volumes of water as well as shrink and swell enormously. Whenever doubt exists as to the way clay will react or is reacting to stress, then laboratory tests should be conducted to verify its real capabilities as a foundation-supporting material.

B. Foundation Investigations

Foundation investigations for new structure or structures requiring major foundation repairs should be made by specialists in this field. Reports covering these investigations should be made part of the structure's permanent record.

On occasion, particularly where spread footings are used, it is necessary to drill borings to check foundation conditions shown by the foundation investigation report. If different conditions are encountered, a new foundation investigation should be ordered. The AASHTO Manual on Foundation Investigations [1], is an excellent guide in both conducting and interpreting foundation investigations. If borings are not of sufficient number or depth, very little knowledge can be gained. For example, a bridge near Cornwall, Ontario, Canada, failed, entailing the loss of 15 lives, as the result of a pier having been founded on a boulder occurring in a layer of "hardpan." The rock had not previously been explored by borings and proved to be only about 2 ft thick. The hardpan was scoured out in the vicinity of one of the bridge piers, exposing clay beneath. The pier eventually tipped over and dropped two of the bridge spans into the St. Lawrence River.

A homogeneous (uniform) soil mass is a soil mass having uniform properties or at least fairly uniform properties. Each layer has essentially the same thickness and is approximately horizontal. A boring at any one point or another will get virtually the same information.

In contrast, an erratic soil mass has soil layers that are not uniform, nor are they consistent in properties, elevation, thickness, or extent. This type of distribution is found more frequently than the homogeneous soil mass. Borings taken at different locations will have different results. Investigations of erratic soil deposits generally try to determine the location and extent of the weaker layers such as those containing silt and soft clay.

C. Physical Properties of Soil

Soils have been classified by a Unified Soil Classification System. Classification is based upon particle size, distribution of the particle sizes, and the properties of the fine-grained portion. Major categories are coarse-grained, fine-grained, and highly organic. Further subdivisions are by grain size, and then by either gradation or plasticity characteristics.

Coarse-grained soils have a single-grained structure. That is, the major force holding the particles together is gravity. There is little attraction between the particles, and they simply rest on top of one another. Large particles have large spaces between one another.

Sometimes in finer silts and clays, chains of particles are formed resembling a honeycombed structure. There is considerably greater attraction between these fine silt and clay particles than there is between coarse particles such as gravel and cobblestones. Here the sizes of spaces between the particles are greater than the particle size.

Very fine-grained soils tend to form a chained structure with voids a great many times larger than the particle size. Most soils are not uniform. That is, they are not made up of particles of equal size. However, a particlar layer is often composed of one general size and type of particle.

Practially all soils contain some water in the spaces between the particles. This stems from the fact that water in some degree is generally present beneath the surface. The soil tends to hold water because of the attraction between the water molecules and the soil particles. Even when soil is dried in the air, the particles of soil, themselves, tend to keep a thin film of water around them. The only way to remove this film is to oven-dry the soil. When all the spaces between particles are filled with water, the soil is then said to be saturated and the degree of saturation is 100%.

Soil particles tend to resist any force that tries to move them. This resistance is simply friction which is developed when two surfaces slide across each other. Frictional resistance of the soil is one of the factors accounting for the shear strength of a soil.

Shear strength is one of the major structural properties of soil. The inspector should ask how well does the soil hold together or how well will it support a structure and permit slopes to be stable? Water up to a certain point has no shear strength and in combination with cohesive soil, reduces the soil's shear strength. To determine the actual shear strength of a soil, an undisturbed sample of it is usually taken to a soils laboratory and tested.

Another important factor in assessing the shear strength of a soil, is its cohesion or the attraction of the soil particles for each other. Cohesion tends to hold the soil together. The amount of cohesion a soil has is usually determined by the minerals that it contains rather than by particle size.

Soils having honeycombed or chained structures show less frictional resistance than soils having single-grained structures. Clay particles tend to be long and plate-shaped rather than angular or rounded as are most silts, sands, and gravels. Cohesive soils are known for the ability to withstand larger strains without rupturing. This characteristic is called plasticity. Clay has a great degree of plasticity. The determining factor in the plasticity of a soil is again water content. A claylike substance, if mixed with a sufficient amount of water, can be deformed freely without breaking up. However, in this fluid-like condition soil has little or no shear strength. If water is removed, the volume of the soil will decrease and eventually the soil will begin to show shear strength. Shear strength is reduced as the moisture content increases. The liquid limit indicates the moisture content beyond which there is no measurable shearing strength.

The inspector's particular concern is with the action of the various types of soil when placed under loads. Loads refer to the pressure exerted upon the ground by the weight of the bridge piers and abutments. The concern now is whether there is going to be settlement or the sinking of the bridge structure due to the consolidation or compression of the sub-soil layers.

Settlement of a bridge pier or abutment may occur from one of the following causes:

Increased loads

Displacement of part of the bed by scour

Lateral displacement of the foundation bed due to lack of restraint

Consolidation of the underlying material

Failure of an underlying soil layer

1. Settlement

Anytime a bridge is built, some settlement is considered inevitable. Through the analysis of soil samples, preliminary field investigations, and the use of some calculations the design engineer can fairly well determine the amount of settlement that will occur. Most of the time rock, gravel, and coarse sand make good foundation materials. However, at times they may have irregular surfaces with varying depths of plastic or unstable soils laid over or under them.

Unequal or differential settlement can have a more serious effect on certain types of bridges such as continuous spans, vertical-life bridges, and fixed arches than on simple spans or cantilevers. An accurate picture of foundation-bed conditions and a knowledge of how certain soil materials might behave under loads allow for fairly accurate predictions of settlement.

In granular soils such as gravel, coarse sand, and medium sand, the maximum settlement occurs as soon as the load is applied, and any settlement thereafter is usually very small. Silt sometimes reacts in a similar way, but settlement may increase somewhat over long periods of time because of a drop in the water table. Unless a soil is confined laterally, saturation will permit serious movement of the soil under pressures of a surface load.

Clays compact somewhat under a load but because of their tendancy to be plastic, they consolidate slowly and at a decreasing rate as time passes. This behavior is due to the slow squeezing out of the water in the clay. Muck and mud that have become trapped under a fill or a firmer layer of soil create a very dangerous situation. Just a small change in pressure can cause this muck to move a great distance resulting in both settlement under the load and heaving in other nearby areas.

An interesting example of failure due to settlement is furnished by the failure of a highway bridge over the LaSalle River at St. Norbert, Manitoba, Canada. The bridge was a single, reinforced concrete arch, with a clear span of 100 ft, the spandrels being earth filled.

The roadway was about 30 ft above the bottom of the river, the height of the fill placed in each approach was about 20 ft. The bridge abutments were founded on piles, driven into the stiff, blue clay exposed at the site and thought to overlie limestone bedrock, based on preliminary borings and the record of an adjacent well. Failure resulted from excessive settlement that caused the north abutment to drop 4 ft and bearing piles to become bent and broken. Subsequent investigations disclosed a layer of "slippery white mud" about 25 ft below the original surface, too insubstantial to carry the superimposed load. Local soils are sediments from an ancient glacial lake and usually overlie compacted glacial till, under which is limestone carrying subartesian water. The existence of this water complicated the underpinning of this bridge foundation, but the work was successfully completed, and the bridge was restored to use. The existence of the white mud in the vicinity was previously unknown; the incident illustrates the unpredictability of glacial deposition. It has a special interest for engineers in that, although the bearing piles were driven into the compacted till (hardpan), settlement of the abutment occurred anyway as the result of underlying soft material.

The undermining of bridge piers by the scouring action of flowing water is one of the most serious threats to bridge foundations. Sudden floods can be extremely destructive due to the great, uncontrollable forces produced by large volumes of water moving rapidly. The removal of soil from the stream bed as well as the damaging effects of scour on the pilings and footings can ultimately lead to the settlement or failure of piers and abutments.

Where floods are known to occur frequently, as is true of the LaSalle River region, the maximum loads and stresses produced on the bridge should be taken into consideration in the original design. Here is a case where this was not done.

Adequacy of the support material for the foundation is not simply the concern of those engaged in constructing a new bridge; it is often a very important consideration when inspecting an existing bridge. Today many of the older bridges are no longer supporting the loads they were once designed for, due to the increased traffic which they must bear. Thus, some of them may be operating under considerably lower safety factors than those calculated at the time of initial construction.

2. Bearing Capacity

Every bridge pier and abutment transmits the weight of the bridge to the underlying ground. The ability of the soil then to support this load, the weight of the bridge and all its parts, is called its bearing capacity. When the load of the bridge is placed on the ground, stresses are created in the soil near the foundation. The greater the load, generally speaking, the greater will be the stress placed on the soil.

Normally, in the case of spread footings the soil lying directly below the base of the footing should be investigated. Coarse sand that is compacted makes a good bridge-pier foundation. Fine sand that is compacted also makes a good foundation material, but it must be confined and not allowed to spread. Spread footings are generally not used where soft soil layers are known to exist at shallow depths. They are commonly used where firm layers are present close to the surface of the ground.

The bearing capacity of any soil depends upon the physical properties of the soil and on the manner in which the load is applied. Bearing capacity is also affected by the type, size, spacing, and depth of the footings and piles. In addition, bearing capacity and settlement are closely related, for in any bridge there is a limit to the amount of settlement that can be allowed. A sudden increase in the water content of a soil layer reduces the frictional resistance and the cohesion of the soil. This, in turn, reduces the bearing capacity and shearing strength of the soils, thereby increasing the possibility of settlement or sliding.

When field investigations show that the foundation-supporting soil layers have a poor bearing capacity, consideration should be given to artificial methods for consolidating the soil to improve its bearing capacity or replacing existing poor soil with material of better bearing capacity. In 1538, an Englishman, who was experiencing severe problems with his bridge settling into the sandy ground, successfully used packs of wool to improve bearing capacity.

Pressure distribution varies between single piles and groups of piles. By producing a lower pressure bulb, piles, in effect, reduce settlement; settlement normally decreases as the foundation depth is increased. However, piles sometimes can cause problems. For example, if you have a fairly thick layer of firm clay located over a deep bed of soft clay, it might be wise not to puncture this firm layer of clay. In other words, a pile might transfer some or all of the load to the soft clay, which is a weaker supporting material, and thus increase the possibility of a foundation failure.

Skin friction indicates the adhesive or shear strength developed between the pile and the soil. Dense granular soils make good foundation materials but piles driven through loose or soft deposits (loose sand, soft clay) have to be driven deep to establish an adequate bearing capacity. Loose granular materials are also subject to large settlements

Cohesive soils such as mud or soft clay do consolidate whereas sands do not. When piles are driven into clays the displacement of clay and such material by the piles often causes the ground to heave. It may cause piles nearby to rise, and if this occurs, reseating of the piles may be necessary.

Groundwater seeping down through the fill in the back of an abutment or retaining wall can weaken the structure of the foundation-supporting soil and cause a shear failure. Once the soil is weakened, it becomes more plastic and tends to flow powered by the force of gravity. A structural element, such as an abutment or a retaining wall, with improper drainage will be susceptible to overturning due to this flow of groundwater.

3. Detecting Foundation Problems

With this background information in mind, the inspector should consider how to go about detecting any changes in the underlying materials.

When inspecting a bridge to determine if foundation movement has occurred, it is sometimes desirable to have certain information about the history of that bridge and pertinent features of the surrounding terrain that might influence the stability of the bridge. One of the best sources of information is the bridge engineer who designed or constructed the bridge if he is still around and available. Assuming he is available, he may be able to provide a description of the foundation conditions at the time of original construction. Another good source is other bridge maintenance personnel and inspectors who have checked the bridge in the past.

A valuable source of information about the soil and conditions in the area may be the local soil conservation office. If soil samples have been taken, tested, and evaluated, then maybe the soil-testing lab at a local university or other facility might have on record data concerning the soil profile near the bridge site or any changes in the bearing capacity of the soil.

All pertinent conclusions of the people or records consulted should be noted, so that they can be reviewed later. For the purpose of inspection, extraneous data such as mathematical formulas and calculations need not be included. In other words, the inspector's notes should be logical and easy to interpret. Simple sketches or photographs might also be desirable.

II. Abutments

Abutments are part of the substructure of a bridge. They provide the end support for the bridge and help retain the approach embankment. There are many types of abutments, and various materials are used in their construction. This discussion will cover the necessary information required by a bridge inspector to properly evaluate and report upon the condition of bridge abutments during his inspection. It deals with the function, types and construction of abutments. Terminology applicable to abutments will be introduced and explained as necessary. The bridge inspector may know the terminology, function, types, and construction of abutments; however, his job is simplified if he anticipates what to look for during an inspection.

A. Types of Abutments

Three common types of abutments are:

Full-height, or closed

Stub, semistub, or shelf

Spill-through, or open

Full-height and stub abutments may be recognized by their location. That is, a full-height will extend from the guideline of the roadway or waterway below, to that of the road overhead. A stub will be located within the topmost portion of the end of an embankment or slope. In the case of full-height, the inspector will see more of the breastwall than in the case of the stub. Abutments may be further typed by their method of construction, such as gravity counterfort, cantilever, etc.

Most abutments are constructed of plain concrete, reinforced concrete, stone masonry, or a combination of concrete and stone masonry. Additionally, some are constructed of steel or timber. Abutments may also be placed on piling.

Most new construction uses the stub, semi-stub, or shelf abutment. Examples of this type of abutment are shown on pages 5-26 and 5-28 of the DOT Manual [2]. Popularly this type of abutment is supported on piles driven through the embankment.

Spill-through or open abutments rest on a column or columns, and as can be imagined, present settlement and erosion problems. This type of abutment, though still used, is not as popular with today's bridge designers and builders.

B. Abutment Problems

The figure on pages 5-25 of the DOT Manual shows an excellent example of the failure of an abutment to perform its function of a bridge-end support. Most inspectors would have recognized it and probably would have reported it as a critical condition. But how many would have suspected that scour caused the condition? On the other hand, which is more important: To recognize abutment failure or to try to determine the cause? The point to keep in mind is that the abutment is tilting out of plumb to such an extent that the bridge appears ready to collapse. It may be nice to know specific causes for each condition observed. However, it is more important to identify accurately all observable conditions so that a more expert judgement as to causes can be made using technical analysis.

As a result of a thorough inspection during which all observable conditions are documented, the inspector may very well determine that scour has occurred. Recall that scour is one of the causes of structure distress when the structure is subject to the flow of water. Consequently, in this particular case the inspector should also have determined and reported the degree of scour.

Abutment problems can be categorized from past experience into rotational movement or tipping, sliding or lateral movement, settlement or vertical movement, or failure of abutment materials. Each of these problem areas are discussed below in terms concerning the bridge inspector.

The visual effect of rotational movement is that the abutment wall is not vertical. The cause or causes of the movement may be scouring, backfill saturated with water, erosion of backfill alongside of the abutment, or improper design. The inspector may identify the problem using a plumb bob, checking clearance between beams and backwall, noting plugged drains and observing cracks. The documentation needed is generally a photograph of particular problem areas, a recording of the results of a plumb bob test, a record of the clearance and a crack measurements, reference to their inspection results such as scour, block weep holes, etc., and a rating (good, fair, poor, or critical).

Lateral movement results from sliding of the abutment. The cause may be slope failure, seepage, changes in soil characteristics, or time consolidation of the original soil. Inspection should include a check of the general alignment, bearings for signs of latěral displacement, construction joints between wingwall and abutment, the approach end joint, and plugged drains. Documentation should again include photographs as well as notation of the direction and extent of bearing movement, of location and amount of separation between wingwall and abutment, reference to other inspection results, and a rating.

Vertical movement is the result of settlement or expansion at the abutment. The causes include soil-bearing failure, soil consolidation, scour, deterioration, cracks, insect attack, or fungus attack. The inspector should check the joint between the approach and deck, cracks, overall superstructure condition, scour, separation between wing wall and abutment, insect attack or deterioration of exposed piling, and fungus attack. The documentation should indicate the type of settlement (even or uneven), the location and measurement of cracks and separation, measurement of scour, photographs, and a rating.

Failure of abutment materials are a result of defects in concrete, masonry, steel, or timber. The causes of the defects are debris, standing water, bridge drainage, mortar cracks, missing stones, insect attack, scour, and fungus attack. The inspector should check for cracks, spalling, rust, rot, drain holes, scour, collision damage, insect attack, and fungus attack. Documentation of the inspection includes photographs; notation of depth, location, and length of cracks; rust; rot; spalling; scour; insect attack; clogged weep holes and collision damage; reference to other inspection results, such as bridge drainage, errosion of embankment and utility leaks; and finally, a rating.

III. Dolphins and Fenders

A. Types

The following discussion will describe the function of dolphins and fenders and what to expect in the way of deterioration or damage to these systems.

The dolphins are those clusters of piles with protective caps. A timber pile cluster dolphin consists of a group or cluster of piles positioned in one or two circles about a center pile and drawn together at their top ends.

Normally the piles are jointed together at their top ends around a center pile with wire rope. In some cases, bolts may be used. Size 5/8 or 1/2 in. wire rope is commonly used and is held in place by clamps.

Dolphins of steel tube clusters or steel sheet piles may also be encountered. A caisson dolphin normally will be constructed of steel sheet piling. It is circular in appearance and may be filled with soil, sand, or similar material. In some cases, the top is covered with a concrete slab and steel or timber wales are attached to the outside.

The function of bridge protective systems is to protect bridge elements against damage from waterborne traffic. Of all the systems available, the most important is the fender system. Many fendering systems have been designed, and many have been analyzed. There are numerous factors to be considered in the design of fendering systems, including the size, contours, speed and direction of approach of the ships using the facility, the wind and tidal current conditions expected during a ship's maneuvers and while tied up to the berth; and the rigidity and energy-absorbing characteristic of the fendering system. The final design selected for the fender system will generally evolve after setting somewhat arbitrary limits for the values of some of these factors as well as reviewing the relative costs of initial construction versus the costs of maintenance. In other words, it will become necessary to make a decision regarding the most severe docking strain or approach pressures one is willing to provide protection against and then design accordingly; hence, any situation that imposes conditions which are more critical than the established maximum would be considered in the realm of accidents and would probably incur damage to the dock, fendering system, or the ship.

It is not surprising that the many and sundry fendering systems an inspector encounters will vary considerably in materials, design, fabrication, and cost. Basically seven types of fendering systems exist today: floating fenders or camels, the standard pile-fender system (timber, steel or concrete), the retractable fender system, the rubber fender system, the gravity-type fender system, the hydraulic and hydraulic-pneumatic fender system, and the spring type fender system.

The floating fender or camel is the simplest type of fendering system employed. This type of fendering system consists of horizontal and/or vertical timber members bolted to the wharf structure. The vertical members may or may not be driven as piles. This type of fendering system was applicable prior to the 1930s. However, with the advent of the larger merchant vessels, particularly the bulk carriers, and with the construction of docking facilities in relatively exposed locations, this system has become outdated.

The timber-pile system employs piles driven along a wharf-face bottom. Pile tops may be unsupported laterally or supported at various degrees of fixity by means of walers and chocks. Single or multiple row walers may be used, depending on pile length and on tidal variation. Impact energy upon a timber fender pile is absorbed by deflection and the limited compression of the pile. Timber piles are abundant and have a low initial cost. They are susceptible to mechanical damage and biological deterioration. Once this happens the energy-absorption capacity declines and a high maintenance or repair cost results. The hung timber system consists of timber members fastened rigidly to the face of a dock. A contact frame is formed that distributes impact loads, but its energy-absorption capacity is limited, and it is unsuitable for locations with significant tide and current effects. The hung-timber system has a low initial cost and is a lesser biodeterioration hazard than the standard timber pile. Steel fender piles are occasionally used in water depths greater than 40 ft, or for locations where very high strength is required and a difficult sea floor condition exists. Its main disadvantages are high cost and its vulnerability to corrosion. Regular, reinforced concrete piles are not satisfactory because of their limited internal strain-energy capacity; also, the steel reinforcement may corrode due to concrete cracking. Prestressed concrete piles with rubber buffers at deck level have been used. In this case, the rubber units are the principal energy-absorbing media, and not the piles. This system is very resistant to natural and biological deterioration.

Retractable fendering systems consist of vertical-contact posts connected by rows of walers and chocks. Contact posts are normally spaced 8 ft on center. The interval between walers is dependent on local tide range. Walers are fastened to holding parts suspended by pins from specially designed brackets. The fender retracts under impact, thus absorbing energy by action of gravity and friction. Energy absorption capacity depends directly on the effective weights, the angle of inclination of the supporting brackets, and the maximum amount of retraction of the system.

In designing this system, tide effect on weight reduction of the fender frame should be considered. Supporting brackets of laminated composite material and proper selection of maximum retraction are feasible means of attaining design capacity. Fenders are more easily removed from open pin brackets than from slot type. In construction, the supporting brackets should be adequately anchored to the associated berthing structure. Although retractable fenders have a high initial cost they have a low maintenance cost, with minimum time loss during replacement.

The rubber-in-compression system functions in the form of a series of rubber cylindrical or rectangular tubes installed behind standard fender piles or behind hung-type fenders. Energy absorption is achieved by compression of the rubber. Absorption capacity depends on the size and capacity of the buffers. In design, a proper bearing timber-frame is required for transmission of impact forces from ship to pier. Draped rubber tubes hanging from solid wharf bulkheads may be used; however, this solid wall should be at least 6 ft in depth, since it is desirable to spread the load over at least a 3 ft height of the ship's hull. Energy absorption capacity of such a system can be varied by using the tubes in single or double layers, or by varying tube sizes. The energy absorption of a solid rubber cylindrical tube is nearly directly proportional to the ship's force until the deflection equals approximately one-half the external diameter; after that, the force increases much more rapidly than the absorption of energy. Rubber-in-shear (Raykin) consists of a series of rubber pads bonded between steel plates to form a series of "sandwiches" mounted firmly as buffers between a pile-fender system and pier. Two types of mounting units are available, that are capable of absorbing 100% of the energy. The only problems with rubber-in-shear fenders are that they tend to be too stiff for small vessels, and the steel plates have a tendency to corrode. Therefore, they have high energy absorbing capacity for larger ships but can damage smaller ships. The Lord Flexible fender consists of an arch-shaped rubber block bonded between two end steel plates. It can be installed on open bulkhead type piers and on dolphins, or incorporated with standard piles or the hung system. Impact energy is absorbed by bending and compression of the arch-shaped rubber column. With the Lord Flexible fender, possible destruction of the bond between the steel plates and rubber may occur. Rubber-in-torsion fenders consist of a combination of rubber and steel fabricated in a cone-shaped, compact, bumper form, molded into a specially cast steel frame, and bonded to the steel. It absorbs energy by torsion, compression, shear, and tension. The main disadvantage here is possible destruction of the bond between steel castings and the rubber.

The final category of rubber fender system is the pneumatic fenders. These are pressurized, airtight rubber devices designed to absorb energy by compression of air inside a rubber envelope. Pneumatic fenders are not applicable to fixed dock-fender systems, but are feasible for use as ship fenders or shock absorbers on floating fender systems. A proved fender of this type is the pneumatic tire-wheel fender which consists of pneumatic tires and wheels capable of rotating freely around a fixed or floating axis. The fixed unit is designed for incorporation in concrete bulkheads. The floating unit may consist of two to five tires. Energy-absorption capacity and resistance load depend on the size and number of tires used as well as the air pressure within the tire.

Gravity fenders are normally made of concrete blocks and are suspended from heavily constructed wharf decks. Impact energy is absorbed by moving and lifting the heavy concrete blocks. High-energy absorption is achieved through travel of the weights. Movement may be accomplished by a system of cables and sheaves, a pendulum, trunions, or by an inclined plane. The main disadvantages of this system are the high initial cost and the high maintenance cost.

The Dashpot hydraulic fender system consists of a cylinder full of oil or other fluid so arranged that when a plunger is depressed by impact, the fluid is displaced through a non-variable or variable orifice into a reservoir at higher elevation. When ship impact is released, the high pressure inside the cylinder forces the plunger back to its original position and the fluid flows back into the cylinder by gravity. This system is most commonly used where severe wind, wave, swell, and current conditions exist. Its main disadvantages are high initial cost and high maintenance and repair cost. The hydraulic-pneumatic floating fender consists of a floating rubber envelope filled with water or water and air, which absorbs energy by viscous resistance or by air compression (or by both means). This fender seems to meet certain requirements of the ideal fender, but is considered to be expensive in combined initial and maintenance cost.

The steel spring fendering system is self-explanatory. Its main disadvantage is the eventual corrosion of the steel.

B. Fendering Materials

The several types of fender systems discussed briefly all involve the use of timber in conjunction with other materials. The timber in each case is intended to absorb a certain amount of impact energy from docking ships and to function as a rubbing surface between the ships and dock.

Accordingly, the timber selected for fender use should have a relatively high compressive strength perpendicular to the grain to resist crushing action. Also, the wood should have a relatively high fiber hardness to resist the rubbing action, although this hardness should not be so great as to result in brittleness and checking. In some fender systems, the timbers are often subject to sizeable bending stresses, in which case the bending strength of the wood should, of course, be relatively high.

Aside from the structural strength requirements of the timber, the presence of marine borers in a particular area should be seriously taken into account. Very severe marine borer activity necessitates the use of treated woods.

Another material of importance in fendering systems is the concrete used in gravity fenders. The materials for making the concrete should be of high quality, providing sufficient strength in the material for the finished product to last an indefinite period of time. Usually the specifications for gravity fenders require 3000 to 5000 pounds per square inch (psi) concrete. The concrete should be made with a high sulfate-resistant cement and should be protected against salt scaling.

Another important fender material is rubber. Most fendering systems utilize rubber in one way or another; as is evidenced by the many companies manufacturing rubber fenders or fender parts.

The final material used to some extent for fendering systems is structural steel. Finding out whether or not structural steel has been used for reinforcement or in steel springs is of importance to the inspector due to the corrosive properties of the steel. Structural grade and durability varies considerably, and these values should have been known prior to induction into the fendering system. Sources of information may be found or obtained from the American Iron and Steel Institute in Washington, D.C. and American Society of Civil Engineers in New York City.

C. Deterioration and Damage

To determine the factors that cause deterioration or damage, keep in mind that these protective systems are normally located to protect the bridge elements nearest the channel. Generally the channel is considered to be the more navigable portion of the waterway. Therefore, damage will most often occur as a result of waterborne traffic collisions. Material deterioration will primarily be of the type caused by exposure to water and the atmosphere near water with the splash zone being the most critical.

Timber deterioration takes the form of decay from fungi, attack by marine vermin, and weathering. Steel deterioration takes the form of corrosion; where steel facing has been scraped by watercraft, corrosion will develop more rapidly. Concrete deterioration takes the form of that caused by water attack. That is, the concrete will probably weather more rapidly or allow penetration by moisture to reach any reinforcing steel, thereby producing corrosion, expansion of the reinforcing steel, and subsequent cracking of concrete.

D. Inspection of Dolphins and Fenders

Inspect timber structures for missing members of the system, collision damage, and vermin attack on all members above and below the water line. Inspect steel members around the splash zone area above and below the water line for signs of corrosion. Check for collision damage such as dents or buckling and for the need for paint. Inspect concrete for spalling and cracking. Check at the water line for weathering of material—the hourglass look.

Also check the condition of secondary elements such as wire rope and clamps. Protective treatment may need replacing. Bleached timber may often provide a clue that the timber needs replacement.

Check the condition of catwalks. Although navigation lights are not discussed here check the presence and condition of required lighting, and finally, check scour, if so required.

IV. Piers and Bents

This discussion deals with the inspection requirements for the piers and bents of bridges. It covers the function, types, and composition of piers and bents. The location and cause of the more common conditions of deterioration or distress is covered.

The function of bridge piers and bents is to support the bridge spans with minimum obstruction to the flow of traffic or water. There are some obvious forces which we can expect will act on a pier or bent. There are other forces which are not too obvious. These loads include:

The weight of the ends of the two bridge spans plus the live load resting on the pier or bent.

Wind loads—which includes not only the wind on the pier or bent, but also the wind on the superstructure and any live loads (moving) thereon.

Traction loads—i.e., the force produced by live loads. A truck braking.

Impact forces—collision damage (vehicular or marine), water debris, and ice.

Pressure of water flow (current), ice (also ice jams), waves breaking on pier or bent.

Buoyancy of the water acting on the pier.

Weight of pier or bent alone.

Movement of bridge spans as they expand and contract (breathe).

Malfunctioning bearings may cause longitudinal forces on piers and bents.

All of these forces are taken into consideration in the design of the pier or bent. For example, if water flow is rapid, it should be obvious that a flat-ended pier will have to contain more or heavier material to meet this force than one with a design that takes into consideration the velocity of the water.

Therefore, when a pier is located in water, the bridge superstructure may be skewed to the pier or vice versa, because the pier has been placed so as to present the least resistance to the flow of water.

If piling is definitely part of the structure, then add pile to the terms pier and bent. In some cases a bridge will have both piers and bents. This should be an easy case for identification, because when seen together the piers will look like piers in comparison to the bents. In any event, when it comes to writing the report, if there is any confusion about the name of an element, a sketch of that element should be sufficient to clarify the matter.

V. Caps

Caps form the top of the substructure. This discussion will consider the subject of caps as they pertain to piers and bents. As discussed previously, the cap is sometimes a difficult item to reach for proper inspection. Inspectors often find many piers and bents situated in water and of significant height. The inspector should be able to identify different types of bridge caps and recognize the difference between good, fair, poor or critical cap conditions of deterioration or distress.

A pier or bent cap functions to properly transfer loads from the bridge superstructure through the substructure to the foundation-supporting material. In simpler terms this transfer takes place as follows. The superstructure loads are transmitted to the cap by bearings. From the cap the loads are passed down through the pier or bent to the footings, which in turn transmit the load to the foundation-supporting material.

Caps for piers and bents may be composed of concrete, steel, or timber. Of these three types, steel is probably the least likely to be encountered. Quite often an inspector will mistakenly identify an end floor beam as a steel cap. Remember that steel pier caps are a special type of floor beam. A close look at the structure should disclose the location of the cap. The function of a cap is to properly transfer loads from the superstructure. If bearings transmit loads to the caps then it follows that caps will normally be located below the bearings.

Concrete caps will generally be reinforced with steel. It has already been pointed out that there are many forms of deterioration or distress that can occur in reinforced concrete.

Timber caps are often encountered during the inspection of timber piers and bents. Sometimes the inspector may find a timber cap on a steel or concrete bent. When inspecting a timber cap, look for those forms of deterioration normally associated with timber.

A pier cap is the topmost portion of a pier. A bent cap is a cap that is the topmost portion of a bent, frame bent or pile bent that serves to distribute the loads upon the columns or piles and to hold them in their relative positions.

The superstructure loads are distributed by the topmost portion—caps—of piers and bents to the foundation-supporting material. From this and the fact that bearings pass these loads to the caps, it can be concluded that caps are subjected to concentrated loads where the bearings are positioned. In design, engineers calculate and provide for an adequate depth of the cap to take care of these loads. However, these points generally call for rather detailed visual inspection. As for reinforced concrete, the areas around the bearing points may be cracked or spalled. These conditions may be the result of poor design or construction error. For example, perhaps the area under the bearing is not sufficient to properly distribute the superstructure loads, and as a result the loads, being "closed in" so to speak by localized compression, tend to cause the concrete to shear or spall. On the other hand, perhaps in the construction phase, the top of the cap was cast so that water tends to stay rather than runoff. This water can penetrate the concrete to reach the reinforcing steel, causing it to rust, expand, and crack the concrete. In the case of timber, more crushing of the cap between the bearing point and pile or column point may result. Water will tend to collect at these points resulting in eventual deterioration of the timber.

VI. Underwater Investigation

The inspection of piers and foundations which are below normal water levels is an operation requiring the combined skill and efforts of both the bridge inspector and a diver. It is a more specialized operation than a routine bridge inspection because a fair amount of sophisticated equipment is required as well as a high degree of skill and knowledge. Therefore, the bridge inspector and the diver must be able to act as a team in relaying the proper information to each other and coordinating the whole inspection procedure.

Not only should the bridge substructure be inspected but also an investigation of the actual stream bed conditions should be conducted. This inspection, coupled with soundings over the whole water area adjacent to the bridge, if performed regularly, can provide a constant check on stream bed conditions.

Underwater investigations should be conducted on a regular basis and after floods. They should include surveys of any structures built in the waterway which are near enough to affect the existing bridge.

A. Inspection of Piling

All too often the piles making up the bridge foundation are forgotten. Since they are usually buried in the soil and cannot be readily inspected, they seldom receive the attention they require. Footer pedestal piles can be seen, but only in cases of scour can seal pier piles be seen. Usually if deterioration or damage is detected before it progresses too far, it can be corrected.

Naturally, inspections of marine piling can be made above the low water line. This is sometimes considered sufficient for small, inexpensive structures that do not regularly bear heavy loads. Some states, such as Florida, require that all structures in water be inspected every two years. Inspection by sending a diver underwater is recommended for bridge piers.

The Navy has developed a rather unique device for inspection of under water structures. The device resembles an ordinary downspout except that it is equipped with a mirror and a light to form a kind of periscope. This device, however, is only effective in relatively clear water at shallow depths. No measurement can be taken with this device. This device, however, has been useful in a number of underwater inspections. It has the one distinct advantage of not needing a diver; for a diver may not always be available, and sometimes the necessary funds to employ a diver fulltime are simply not available.

1. Wood Piling

Due to the effects of salinity changes on the life of some marine borers, seasonal changes in salinity should be noted and watched. Increases or decreases in the salinity content may result from the inflow of fresh water from nearby streams and rivers. Changes in salinity may depend on the depth of the water so that general conclusions should not be made from results at any particular depth.

Many times it is very difficult to inspect all piles because of the large number. Thus, it may only be necessary to inspect a representative number of the piles where underwater conditions are uniform.

The effects of marine borers on wood piles should be checked carefully. Once a pile has been penetrated, it becomes increasingly vulnerable to further damage or deterioration.

2. Steel Piling

Steel generally does not corrode very much in fresh water. However, polluted waters can cause corrosion if the steel piles do not have any protective coating. Piles which have been pulled after 19 years of service, have shown little evidence of corrosion. However, it should not be assumed from this that all steel piling will fare as well. In sea water steel piling tends to corrode much more rapidly and intensively than in fresh water. Certain zones of marine steel piling are subject to more intense corrosion than others.

Brackish water, water that is part fresh and part salt, with little movement, caused the worst problem in Florida. Another area of corrosive action occurs between the unaerated area below the mud line and the aerated water just above the bottom. High water temperatures tend to speed up corrosion and they also speed up the accumulation of marine growths.

Corrosion on H piles underwater is apt to proceed more rapidly on the edges of the flanges, because of differences in oxygen concentrations there and on flat surfaces and because of abrasion. Sometimes scour can cause an increase in corrosion due to the fact that it can break off corroded steel on the surface and expose the inner portion of the pile to corrosion. The outer film of the pile acts as a protective layer for the inner steel.

3. Concrete

Abrasion or scour of underwater concrete structures sometimes results from the action of water flow and debris and sediment in the water. Waves surging against concrete piers, or blows resulting from ships colliding against concrete piers, or blows resulting from ships colliding against the substructure can cause bending of the piers and localized cracks to form. These cracks can permit water to enter the concrete, resulting in corrosion of the reinforcing steel and in spalling. If sea water gets inside the concrete it can cause shrinkage or cracking. Small cracks below the low water level tend to close themselves sufficiently to prevent rusting of the reinforcing steel. Thus, they are not normally detrimental to the reinforcement.

Chemical decomposition of concrete in sea water is also promoted by the existence of cracks in the concrete. Sulfides or acid in the water can also cause deterioration of concrete and the exposure of the steel reinforcement.

In warm, subtropical waters, there is always a possibility of rock borers such as Pholads entering poor concrete. Normally, quality concrete prevents this as well as other problems. Freeze/thaw cycles also tend to break down concrete. Water in hairline cracks expand, thus promoting deterioration. Again, this problem deals directly with quality concrete design and construction technqiues.

B. Underwater Inspection with Divers

Florida's Department of Transportation has been a pioneer in the development and use of underwater closed-circuit television systems for inspecting bridge substructures. The state has been using divers for underwater inspections since the early 1970s.

Normally, the inspection is performed by two divers, one operating the camera while the other assists him in positioning the equipment or in some other task. Heavy marine growths are removed with scrapers prior to using the closed-circuit television system. Both divers are under the direction of an engineer who monitors their activities on a television set located above the surface.

Divers should operate as a team not only from the standpoint of efficiency but also safety. Some danger always exists in diving regardless of depth involved. A safety man should be available at all times. Someone should always look out for dangerous marine life such as sharks, moray eels, or spiny sea urchins. If the current is quite swift, divers should not be working underwater.

As one would expect, underwater inspection requires a great deal of preparation and coordination. First of all, it is necessary to make sure that a qualified diver is available. Then the proper equipment must be obtained through various state and local resources. The diver must be thoroughly briefed about what he is to look for and the type of information he is to record or relay. For example, if foundation settlement is suspected, then the diver would be asked to look for evidence of cracking, especially vertical cracking in the concrete of the piers and footings. The diver should investigate the extent to which any such damage has affected the substructure.

One of the biggest problems encountered is that few divers are qualified bridge inspectors. It is quite possible that the diver will pass over something a bridge inspector would consider to be of vital importance. It is obvious then that the inspector's role is to instruct and inform the diver as thoroughly as possible about what he should be looking for.

Even when a bridge inspection team is using some of the most sophisticated equipment, visibility underwater is often limited by such factors as turbulence from currents, sediment in the water, or the distortion of light as it passes through the water. Many times, it is impossible to see more than 3 to 6 in. There are several types of distortion which are commonly encountered in using underwater cameras. One of these is distortion which forces the diver to move away from the object in order to get all of it in the picture. Therefore, he must shoot through more water and a greater amount of sediment, which causes more light energy to be absorbed and consequently reduces the clarity of the picture. A strobe light will help eliminate this problem. Another type of distortion is spherical distortion that results from changing the focal length from the center of the picture outwards to obtain a wide angle view. This causes a decrease in resolution or clarity and produces a fuzzy haze around the edges. False images or a shadow effect can result from several factors, such as turbulence and light distortion. Sometimes straight lines simply are not straight but, rather, are seen as wavy or curved. Thus aberrations or lack of focus must be eliminated as much as possible in order to obtain a sharp, clear picture.

An approximation of underwater light distortion can be seen if one walks into the hall of morrors at an amusement park fun house. Your body is distorted in a different way by each mirror. A similar effect is sometimes produced by the water, so that images are not always seen in their true perspective.

VII. Culverts

One way of thinking of a culvert is to consider it as a grade-separation structure located between two types of traffic: The water traffic flows through the culvert and the highway traffic flows over it. It is, first of all, essential that a culvert be able to handle the discharge of water flowing through a channel at any particular time. When a culvert ceases to function the resulting flooding can be quite destructive and consequently create quite an expense. In addition, there may be delays in traffic flow and interruptions or even fatal accidents because the culvert is clogged with debris and prohibits the normal flow of water. Should malfunction occur, embankments, roadway, and nearby property can be subjected to extensive damage.

A culvert must also be strong enough to support fairly heavy loads. Not only must it bear the overlying fill but it must be able to withstand the traffic loads passing over as well. Two types of load are present, the fill or static load because it is always present and relatively constant, and the traffic load or dynamic load since it varies from time to time in relation to the size and speed of the vehicles passing over it. Thus, the ability of the culvert to support such loads depends on its strength and how it is bedded and backfilled.

Culverts need to be durable in order to withstand great temperature changes, corrosion, and the abrasive effect of water, sand, gravel, and debris passing through them. Long life and durability also are dependent upon good maintenance and conservation practices.

Metal culverts must be protected against acid water or soils that could accelerate corrosion activity.

Culverts are seemingly insignificant and much-taken-for-granted structures when they function normally. They can, on the other hand, cause much havoc and distress when they fail to function.

A. Location of Culverts

The grade of the culvert commonly follows the approximate line and grade of the natural stream channel. Flow line coincides with the stream bed for the most part. Thus the stream bed offers the best hydraulic alignment even though it might be curved or uneven. The advantages of culverts are lesser if the fill is very high because culverts are long and the static earth loads are huge.

The most important factor in locating a culvert is to insure that it will carry water adequately under the embankment. Placing the culvert in the existing channel generally is the best policy in this respect. Inspectors should note the apparent effect of the structure on both downstream and upstream erosion. Other factors affecting the choice of location are the weight of the surface load which the culvert will be required to support and the foundation conditions.

B. Diversion of the Stream Channel

It is not always practical to place the culvert in the natural channel due to its skew angle with the roadway. Generally a longer culvert would be required if the natural channel is followed. The shortest length of culvert is that which is placed at right angles to the roadway.

Either the inlet or the outlet channel may have been diverted. It is also possible to divert the complete channel, but a new channel must be formed to carry the stream water. When the stream was diverted from its natural channel, steps were taken to minimize scour and erosion at the toe of the embankment. Usually a culvert functions best when the entrance is located in or as near to the natural channel as possible. Special considerations such as cost of protective measures or cost of a longer culvert or right-of-way required for diversion channels usually determine the location of the culvert. Ultimately, a well-designed culvert should minimize scour and the entrapment of debris. The inspector, for his own knowledge, can observe the effects of good or poor design. Such effects, if detremental to the overall function of the culvert, should be reported.

C. Design Discharge

Culverts are designed to accommodate the so-called design discharge or runoff. Several methods exist for determining this discharge value. Rainfall influences runoff and tends to vary geographically. In Louisiana, for example, an average storm lasting an hour may produce about 3 in. of rain. In Southern California less than one in. is normally produced in an hour under similar conditions.

Factors affecting runoff intensity and duration include intensity of rainfall, the character of soil and soil surface (since runoff is increased by soil saturation), vegetation (as plants absorb a certain amount of moisture, and farm crops planted in rows parallel to the direction of flow may increase runoff), surface storage such as lakes and swamps, and topographical features, such as the size of drainage basin and slope. Also, the shape and slope of the main channel have an effect.

Because of the large number of variables, determining the quantity of water a culvert needs to handle often depends to a large degree on experience and good judgement based on a knowledge of local conditions. Through the efforts of the Soil Conservation Service, the Bureau of Public Roads, State Highway Department, the Army Corps of Engineers, and other groups, more and more data is being accumulated on rainfall and runoff. In some areas, sufficient records of runoff are now available to permit good estimates of the requirements of culverts in that area, but many installations still depend on special formulas plus experience and good judgement.

D. Slope

Steep culvert gradients tend to minimize the problem of water backing up but they increase the possibility of erosion at the outlet because of the high flow velocity. Flat culvert gradients tend to cause silt and debris deposits to develop in the barrel. To prevent this sedimentation from occurring, the culvert slope should produce a velocity through the culvert barrel greater than that of the normal stream velocity. Concrete box culverts, which have a low gradient, can have problems of silting and collecting debris; however, those with too steep gradients can have scour and erosion problems.

E. Debris Control

Debris can cause serious problems if they are not trapped some distance upstream of the culvert entrance. Similarly, sand, gravel, and larger rocks can cause abrasion if not deposited before reaching the culvert. Floating debris carried by the stream includes anything from small twigs, pine needles, and garbage to large limbs, logs, and uprooted trees. Floods carry the heaviest loads of debris; sometimes they even move large boulders. If debris is trapped at the culvert proper it should be removed.

Periodic maintenance is required to remove the debris from debris control structures; and thus prevent the upstream channel from being blocked or becoming silted up. The type of debris control structure used depends on the amount and variety of debris in the channel. Heavy debris such as rocks and boulders is common in mountainous areas. Rolling countryside or flat cultivated land usually provides finer types of sediment.

When clogging of a drainage structure by debris is a problem, debris control structures may have several advantages. They (1) prevent traffic delays because of accumulation of drift on the roadway or washouts caused by clogged culverts; (2) allow for planned maintenance during floods rather than during emergency maintenance (this holds true for other emergency situations that exist during floods); (3) provide a "safety factor" by sizing a culvert to accommodate a specific estimated quantity of debris; and (4) supply a safeguard against damange from bouyant forces when an accumulation of debris at the culvert entrances causes partial flow.

Debris barriers are essential to the proper functioning of some culverts. Debris can be controlled by three methods; intercepting the debris at or above the inlet, deflecting the debris for detention near the inlet, and passing the debris through the structure. Debris control structures, if utilized, must be inspected and maintained on a regular basis, to ensure adequate functioning.

F. Inspection of Culverts

When undertaking a culvert inspection, the first task of the inspector is to look at the general condition of the culvert. How is it aligned in relation to the stream channel? Is there debris blocking the entrance? These are the superficial items which are readily visible.

The inspector should determine the condition of concrete including the evaluation of the extent of spalling and abrasion in the box or barrel and at headwalls and endwalls. Notes should be made of the extent of exposure and location of exposed reinforcement. Cracking of concrete should be recorded as to size and length, type (vertical or horizontal), and location. Finally, any other signs of deterioration should be noted.

Abrasions in culverts depend upon the solid materials, such as sand and gravel, that are carried in suspension by the stream. These particles abrade the floor and walls of the culverts as they pass through. The larger particles are the most abrasive, for they tend to be tumbled along the bottom. Check for small holes and imperfections and check the joints to see if they are tight and not separated. Expansion joints can be a serious problem because the filler material will probably be too soft to resist abrasion.

Good quality concrete will be resistant to wear from clear water traveling at high velocities as long as the flow is uniform and the direction of the flow is not abruptly changed. Small holes, imperfections, or misalignment of the floor or sides may lead to serious damage as a result of water turbulence. Streamlining is the best way to prevent or reduce erosion where high velocities of water are common. In general, the erosion resistance of concrete increases as the strength of the concrete is increased.

The deterioration of concrete in culverts is not unusual. Common types of mechanical deterioration are abrasion or erosion, overstress or overloading, shrinkage failure, freezing and thawing, fatigue, and reaction to acid or alkaline soils.

One rates the condition of metal culverts by determining the extent of abrasion, pitting, corrosion, and the condition of rivet and bolt connections.

Erosion of a culvert fill is speeded by a lack of vegetation to hold the soil in place. Heavy rainfall, naturally, increases the runoff of water from the fill, carrying large loads of sediment into the stream, which then contributes to the silting up of the culvert. In addition, large amounts of water seeping into the embankments can produce slide failures which can block the stream channel and the culvert inlet or outlet. Check to see if drainage piping is functioning properly and the fill is being adequately drained.

Water which seeps into the embankments and escapes from poor culvert joints can result in the removal of supporting material along the outside of the culvert, a condition known as piping. If inspection of the culvert ends reveals voids around the culvert periphery or piping, then corrective measures should be initiated. These measures include placement of barriers, such as clay blankets, around the culvert ends and water proofing the culvert joints.

Other inspection items include sudden dips, sags, or cracks in the roadway immediately over the culvert. Look along the roadway barriers for signs of cracking in the concrete. Also check the approaches for signs of sagging or cracking that might also indicate settlement. Check for settlement of culvert structures by looking for sag in the culvert floor, canted wingwalls and cracks in top slab. Cracks in the top slab or wingwalls also provide evidence of settlement. Water seeping through these cracks can cause further deterioration especially if the water freezes and then thaws.

Check condition of culvert joints by looking for vertical differential settlement at the expansion joints, transverse and longitudinal differential settlement at the expansion joints, widely opened expansion joints and water seeping through joints from soil outside.

Observe general overall condition as to the alignment of culvert to water flow, missing or broken sections, flood damage, scour damage, headwalls and endwalls, box or barrel, debris and sediment accumulation, and finally backwater.

Remove debris, earth, and sediment as necessary to facilitate inspection. Sometimes it is physically impossible for the inspector to remove the sediment from a culvert that has experienced heavy silting. It is also difficult to remove large trees that may be blocking the culvert entrance. In such cases, the inspector should complete as much of the inspection as possible and then indicate on the inspection report that maintenance must be implemented by trained personnel before he can complete the inspection. If the culvert is endangered by these deposits of debris, earth, or sediment, then the inspector must emphasize the urgency of the need for its removal.

Check the channel upstream from the culvert for any obstructions such as fallen trees or earth slides blocking the channel. Such situations could lead to heavy sedimentation or even lead to washout if the debris or earth barrier were to collapse.

Note any construction activity or other projects in the vicinity of the culvert that might endanger its structural integrity. Check to see if large quantities of debris or earth are entering the stream channel. Check to see if the culvert is adequate to withstand any heavy loads being placed upon it by noting any sagging or any structural deficiencies in the box or barrel.

Recommendations in the inspection report may include removal of debris or sediment, additional scour protection, construction of debris racks, repaving invert or outlet, more riprap needed, repairs required, or replacement of wingwall. Based upon results of the inspection, the inspector should make any recommendations considered important. If repair or replacement is necessary, then these facts should be documented and supporting information should be included in the report.

VIII. Examples of Substructure Inspection

A. Abutment Ratings

The bridge abutment is subject to numerous types of forces and resulting duress. The abutment is designed to carry dead loads and live loads of the superstructure, soil pressures behind the abutment, and occasionally water pressures from streams or rivers. In addition, quite often forces produced by settlement, "growing" concrete pavements, or shifting of the superstructure are imposed on the abutment. Debris collecting in the cracks of concrete pavements in cold weather prevents expansion of the concrete on hot days and causes the pavement to "grow." The expansion is transferred to the bridge site and often causes tilting of abutments and other damage. Settlement and other forces frequently cause damage to the bridge abutment. Water drainage from the superstructure often carries salt that deteriorates the concrete or it infiltrates the reinforcement and promotes corrosion.

Figure 7.1 shows an abutment in good condition. The abutment in Fig. 7.2 is tilting, and duress is evident from bulging timber. This abutment is rated, poor. The abutment in Fig. 7.3 is in critical condition, and a temporary timber bent has been placed near the abutment to carry most of the superstructure loads. Note the wedges used to maintain proper support.

Fig. 7.1 Abutment in Good Condition

Fig. 7.2 Abutment Rated in Poor Condition Because of Tilting and Evident Duress

Fig. 7.3 Abutment in Critical Condition—Deterioration and Movement Visible

B. Pier Ratings

The bridge pier is subjected to many of the same forces as the abutment: The superstructure loads, settlement, water pressure or ice floes in rivers, debris in rivers, and occasionally, collision damage from vehicles. The pier and pier cap are also subjected to drainage from the superstructure and deterioration similar to that which the abutments experience.

Figure 7.4 shows a concrete pier in good condition. The pier in Fig. 7.5 is in fair condition. Some spalling of the concrete column is evident and the reinforcement is beginning to corrode with exposure to the atmosphere. The fair rating is given since the carrying capacity of the member has not yet been greatly reduced. The concrete pier shown in Fig. 7.6 is in critical condition. The concrete is badly deteriorated, steel corrosion is severe and the steel-and-concrete bond is no longer effective. The load-carrying capacity of the pier has been greatly reduced as a result. The deterioration is apparent.

Fig. 7.4 Concrete Pier in Good Condition

Fig. 7.5 Spalling of Columns Give Pier a Rating of Fair Condition

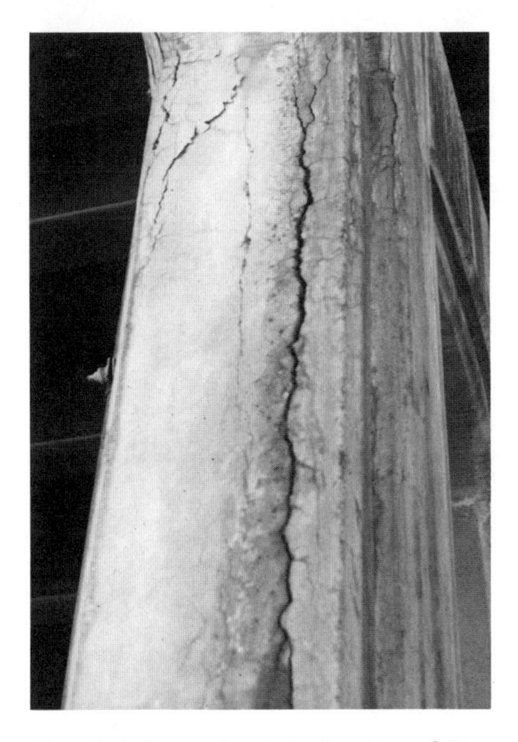

Fig. 7.6 Extensive Deterioration of Concrete Places this Pier in Critical Condition

C. Settlement Indicators

Settlement is a primary cause of bridge damage. The inspector should, therefore, be alert to various indications of settlement. One of the best indicators is the bridge railing. Dips or tilting of the bridge railing is often an indication of settlement. Figure 7.7 is an example of this condition. The poor alignment and damage of the bridge railing is a result of settlement of this bridge.

Tilting or rotation of abutments and piers, misalignment of joints, and movement of associated structures such as approach slabs, curbs, utility lines, or drainage ramps may also indicate settlement has occurred. Figure 7.8 shows cracking and movement of the drainage ramp and approach near the end of the bridge. This movement indicates that settlement of the abutment has probably occurred. The inspector should always be aware of such indicators of duress to the bridge during the field inspection.

Fig. 7.7 Dips and Tilting of Bridge Railing, Often an Indication of Settlement

Fig. 7.8 Movement of Abutment is Often Indicated by Cracks or Movement of Wingwalls and Approach Slabs

References

1. American Association of State Highway and Transportation Officials, Manual on Foundation Investigations, Washington, D.C., 1967.

2. U.S. Department of Transportation, Federal Highway Administration, Bridge Inspector's Training Manual 70, Washington, D.C., 1971.

Chapter 8

MOVABLE BRIDGES

I. Introduction

Movable bridges of narrow span have been constructed worldwide, from a fairly early date. Spans of some size, however, have been constructed only in recent years. In 1850 the design for a vertical lift bridge with a span of 100 ft and a rise of 54 ft to cross the Rhine River at Cologne was presented by Captain W. Morson, an Englishman. This bridge, however, was not built, nor was the bridge designed by Dr. Waddell in 1892 to span the ship canal at Duluth, Minnesota. This latter bridge was to be 250 ft long with a lift of 140 ft.

Probably the first movable bridge of any size to actually be constructed was the South Halstead Street Bridge in Chicago. This bridge, designed by Dr. Waddell in 1892, had a span of 130 ft and a maximum vertical clearance of 155 ft. Since 1908 many well-designed and economical bridges of this type have been built.

II. Types of Movable Bridges

A. Vertical Lift

The vertical lift bridge is simple to design and construct, for the complicating details are comparatively few and present no difficult problems. From the standpoint of actual operation, the vertical lift bridge offers exceptional reliability, ease of operation, and great potential length of span when compared with bascule or swing bridges. The duration in time of the opening and closing cycle of a vertical lift bridge is small compared to a swing bridge and therefore may be kept open to surface traffic longer at approach of a vessel.

The well-designed vertical life bridge offers great economy in its power requirements, which are usually below those of a swing bridge where fast operation is a factor.

From the standpoint of a clear channel, the vertical lift is probably the most satisfactory of any movable type, if the lift is sufficient, for there is essentially no interference whatsoever with the complete breadth of the channel. In the crowded conditions of urban areas the vertical lift bridge, in common with the bascule bridge, offers the additional advantage of occupying so little of the channel area that new spans can be erected adjacent and closely parallel to the old ones.

Most of the vertical lifts have a lifting span and have two towers of four columns each. Towers with inclined rear columns are more economical than those with all columns vertical, and are generally used except in special cases, such as the skew lift bridge; or occasionally they are foregone for the sake of appearance. When the towers are built with four columns each, overhead trusses connecting them are rarely needed unless the bridge is to carry pipes, heavy wires, or cables. Bridges of this type have been built with spans up to 544 ft and with lifts as great as 140 ft and with weights to over 3 million lb.

Vertical lift bridges with two-column towers are usually restricted to short lift spans of moderate height.

A few vertical lift bridges have been built with lifting decks, a style which is sometimes advantageous when two decks are required for the traffic—usually the upper for highway traffic and the lower for railroad passage. Thus, when the overhead span is fixed only part of the land traffic is inconveniences by ship movement; and, in the case of lifting deck and a lifting span, the lifting span needs to be moved only in the case of the passage of very large vessels.

B. Bascule

The earliest form of bascule bridge was undoubtedly a simple span, "trunnioned," or hinged at one end, moving in a vertical plane about the trunnion. Movement was obtained by an out-haul line attached to the free end and running upward and inward to a person or donkey, which was the source of power. This is, of course, merely a description of the familiar medieval drawbridge used to carry traffic over artificial moats and canals. These bridges and some of the earlier modern types were poorly counterweighted, if at all, and thus their widespread utility was limited.

The word, bascule, comes from the French word meaning see-saw; hence a bascule bridge is, more properly, a balanced bridge. The modern bascule bridge is so balanced that the counterpoise, sinking into a tailpit, lowers as the roadway rises. Although a few bascule bridges of some proportion were constructed in Europe during the first half of the nineteenth century, the real beginning of the development of the modern type may be considered to have started about 50 years ago.

1. Advantages of the Bascule Bridge

From the standpoint of design, cost, and operation the bascule bridge is similar to the vertical lift. Its speed of operation, duration of cycle, interference with channel, noninterference with adjacent piers, adaptability to wide roadways, and ill effects of collision with river craft compare so favorably with the vertical lift bridge that a choice between the two types often becomes largely a matter of aesthetics. The relative economy of the bascule in comparison with the vertical lift depends on specific conditions in individual cases. In general, however, the vertical lift is the most economical for long spans and low lifts in localities where only a limited vertical clearance is required.

There is one advantage the bascule has over the entire field. This is the protection of traffic, a consideration which becomes of constantly increasing importance in the busy parts of any city. By locating the break in the roadway floor of a double-leaf bascule bridge ahead of a vertical plane passed through the axis of the trunnion, the leaf of the deck trunnion bascule forms a traffic barrier from the instant that it begins to lift. Such a barrier is continuous throughout the entire time of opening and closing. An automatic safety barrier of this sort could be obtained on other moving bridges only through additional expenditures for barriers and controls.

2. Types of Bascule Bridges

At the present time there are in use four principal types of bascule bridge: (1) the Scherzer type, a rolling lift type; (2) the Rall, a rolling lift type; (3) the simple trunnion, or Chicago type; and (4) the Strauss style—a modification of the simple trunion type.

The simple trunnion, or Chicago type, is the bascule type to which Torrington bearings have been applied. The first example of this type of bridge was the Claybourne Avenue Bridge which was built in Chicago at the very end of the nineteenth century. Throughout the intervening years 40 similar spans, with new refinements incorporated at each step, have been erected by the city of Chicago. The two most recent examples are the Canal Street Bridge and the North State Street Bridge, which were completed in 1948.

Because the busy Chicago River bisects the business district of the city and its north and south branches reach far into the industrial sections, the peculiar advantages of the bascule have been developed to the full by Chicago engineers. Among the considerations which have led to the almost exclusive use of this type the following should be mentioned: (1) the Chicago River is narrow, and bascules require no center pier which might obstruct navigation, (2) a minimum amount of space is required for the approaches and foundations along the crowded banks of the river, (3) a partly open bascule is often sufficient for vessel passage and results in shorter operation time, (4) the nature of the type permits building numerous bridges along the length of the river without interference with adjacent docking space.

Although the actual engineering details of the single-leaf are essentially the same as those for one of the double-leaf trusses, the decision for one or the other arrangement is dependent upon numerous factors. The cost factor may not necessarily differ for the savings which result from the erection of one leaf may be more than offset by the special requirements of the longer single truss and the greater depth of the tailpiece, or counterweight pit.

Speed of operation is, of course, also a matter of great importance in the traffic control of a great metropolis, where the delay of even a few minutes can snarl automobile movement for an hour, or more. Less time is required by the double-leaf style to provide an open center channel than by the single-leaf.

In appearance, a factor of greater importance than is generally conceded, the double-leaf bascule is far superior to the single-leaf. Because the stresses in the trusses of a double-leaf bascule are the greatest at the piers and at a minimum at the center of the channel, a curved-chord cantilever may be used to give the closed bridge the appearance of an arch. This arched effect is not only pleasing to the eye, but saves metal at the center and increases the available headroom at that point, thus permitting a greater number of vessels to pass beneath the closed bridge.

The lighter weight of the shorter leaf of a double-leaf bascule results in several advantages that have made the style a desirable one for Chicago, or for any other location presenting comparable problems. First, the leaves are subjected to less wind pressure than the single leaf of the other style, and therefore the overturning on the pier and the maximum soil pressure are reduced. These are important considerations when the foundations are in soft material or on piling. Second, because the moving leaf of the double-leaf type is shorter and lighter, the counterweight arm may be made smaller, reducing the distance from the grade of the roadway to high water elevation. Third, by cutting the length of the counterweight arm it is possible in many cases to prevent the counterweight from dipping below the water line when the bridge opens, thus effecting considerable savings by eliminating the need for a watertight tail-pit. This is an especially useful characteristic of the double-leaf bascule when the grade of the roadway lies close to the water's surface at the bridge site. Fourth, the reduced weight and moment arm of the overhanging leaf of the double-leaf type makes more economical the counterweight material itself. That is, less material is required for both counterweights on the double-leaf type bascule than is required for the one counterweight on the single-leaf bascule.

C. Swing

It sometimes happens that the physical surroundings, operating conditions, or cost factors are such that it would be impractical to use either the vertical lift or the bascule bridge to solve a movable-span problem. The alternative that is most frequently resorted to, in this case, is the swing bridge.

The swing bridge, which may be either rim-bearing or center-bearing, pivots on its center so that the span, when open, lies parallel to the axis of the river or stream. Each truss of the center-bearing swing bridge is supported at the end of a cross girder which is, in turn, supported by a center bearing. So that the span will not tip when it is open, balance wheels, which follow a circular track on the central pier, are fastened to the trusses and the floor system. These wheels carry no load unless the bridge becomes unbalanced. In this type of swing bridge the dead load of the span is carried, when open, by the cross girder, but when closed the ends rest on the approach piers in such a fashion that the latter support the live load. The trusses of the rim-bearing swing bridge are supported by a circular girder that turns with the span. This circular girder moves on as many conical rollers as its circumference will allow, so that the many bearings thus provided will bring possible span deflection to a minimum.

The swing bridge, although a well-known type, is limited in usefulness. Its greatest drawback is probably one of space, for it requires a sizeable central pier, which often adds appreciably to water-traffic congestion on busy rivers. It is slow in action and requires a large amount of clear river space in which to open. These and other limitations render the type generally inferior in practicality to the vertical lift and bascule bridges.

When none of the foregoing conventional types of movable span bridges has been found suitable for a specific situation, engineering genius and imagination often have been combined to produce unusual, but eminently successful, bridges. Such are the Lake Washington floating bridge and the Borden Avenue retractile bridge.

Pontoon bridges, civil and military, have been widely used since ancient times, but it took daring and imagination to conceive the Lake Washington Bridge as a permanent and major link in the State of Washington's highway system. Once adopted, it took ingenuity to devise the draw pontoon, its basin, and the incredibly large floating sections.

Likewise, the narrow width of Dutch Kills, the restricted grade of the approaches, and the small amount of area available must have seemed, at first, to present an insurmountable problem. The solution, however, by which the bridge span is retracted obliquely to provide a clear channel, seems crystal clear after one sees the completed design.

III. Operation of Movable Bridges

A. Vertical Lift

1. Mechanics

The ordinary vertical lift bridge is simple to design and construct. The complicated details are comparatively few and present no difficult problems. The vertical lift is as rigid as other types of movable bridges and the reliability of the vertical lift has been demonstrated many times. The South Halstead Street Bridge of Chicago in 1907 was said to be the most satisfactorily operating movable bridge in the city.

A well-designed vertical lift is as easy to operate as any other type of movable bridge. The time required for a complete raising or lowering of a vertical lift bridge is usually about 45 to 50 sec. Of course, the time required for a partial raising or lowering is less.

Vehicular and pedestrian traffic over vertical lift bridges may be amply protected during the raising and lowering of the lift span by suitable gates.

A vertical lift bridge seems to have an advantage over bascule and swing bridges in regard to possible boat collisions. With the movable span in place, it appears that a boat could do more damage to a swing bridge and draw rest than to either a bascule or vertical lift bridge. Probably less time would be required to repair a vertical lift span than either a bascule or swing bridge. If the movable span was partly open when hit by a boat, the swing span and draw rest of a swing bridge would probably be crumpled up. In the case of bascule, the bridge would suffer only slightly if struck by the top hamper of the boat, but considerable damage would result if the hull hit the bascule leaf. A vertical lift span partly raised would probably be high enough to damage the mast, rigging, smoke-stacks, and pilot house of a boat, and, consequently, no very serious damage would be done to the bridge.

Thus, it appears that the possibility of serious damage due to a collision with a boat would be less with a vertical lift than with a bascule bridge and less with a bascule than with a swing bridge.

2. Design Features

The following notes are on vertical lift bridges and include information on the particulars in which vertical lift bridges differ from ordinary bridges with fixed spans.

In general the lift span truss is designed in the same way as the ordinary fixed span truss with the exception that suitable seating devices must be provided at the ends of the lift span and means must be devised for fastening the various cables attached to the span. Provision must also be made for the placement of the necessary machinery, machinery house, and operator's house on the lift span.

For long spans and high lifts, each tower should be composed of two vertical front and two inclined rear columns, well braced in both directions. Provision must be made for fastening the sheaves on the vertical columns by suitable sub-posts or by a sheave girder. In this type of tower the counterweights move up and down inside the tower.

For short spans and low lifts, where the lift span is usually a plate girder, the towers should each be constructed of two vertical columns with sway bracing between them. The tops of the columns of the two towers should be connected by light trusses to assist in keeping the towers vertical and to hold them the correct distance apart. Sheaves are placed on top of the columns and the counterweights move up and down outside of the columns.

When the lift span is built on a skew, and the towers are supported independently on masonry foundations, it is desirable to construct each tower of four vertical posts well braced in both directions. Horizontal bracing should be provided near the tops of the towers. The sheaves should be placed on all four columns of each tower and the counterweights at the rear of the towers.

Towers should be cambered so that they will be vertical under dead load only (there should be no live load on the lift span when it is raised and lowered). No camber is necessary for towers having two or four vertical columns resting on masonry and having a sheave on the top of each column. When the tower is composed of two vertical and two inclined columns, with the vertical front columns resting on a supporting truss, the tower should be cambered so that the vertical columns are really vertical when the lift span is being raised and lowered. In this type of tower the sheaves are usually placed directly over the vertical columns, hence these columns carry practically all of the load.

Guides should be attached at the eight corners of the lift span in order to keep it in line while being raised and lowered. Roller guides that fit with sufficient clearances in vertical tracks on the tower are the preferable type. The guides should be designed for wind loads on the span and also for traffic thrust and any other loads that may be applied to them.

Centering blocks should be attached to the four lower corners of the lift span to hold this span in place when it is in its lowered position. These blocks should engage blocks attached to the base of the towers. Provision must be made for some longitudinal movement at one end of the span.

Counterweights for vertical lift bridges should be made of concrete cast on a steel framework. This framework should be strong enough to carry the concrete when attached to the lifting cables. Cast iron blocks may be used when sufficient space for the concrete is not available.

Counterweights should weigh about 5% less than the weights to be balanced. Movable weights equaling 10% of the balanced weights, should be provided so that the proper balance may be obtained between the lifted weights and the counterweights. These movable weights should not weigh more than 200 lb each and they should be provided with eye or U bolts, for handling. Safe places should be provided for these movable weights in the top of the counterweight framework. The space provided must be such that none of the movable weights will project above the top of the counterweight framework.

The inside face of the counterweight should be provided with guides that engage tracks on the tower. Ample clearances should be provided so that the counterweight will not bind or stick. Clearances of 2 in. or more should be provided between the tower steelwork and counterweights. If a counterweight is composed of two or more parts, about 2-in. clearance should be allowed for between the parts. When the counterweight is in its lowest position, it should be not less than 3 ft above the bridge floor.

The counterweight cables for vertical lift spans should preferably be of plow steel and consist of six strands of 19 wires each, constructed around a hemp center. The sockets may be open or closed, but should be of standard design.

When the span is being raised, movement of the lifting and counterweight cables will cause an unbalanced condition between the span and the counterweights. This condition may be met by adding extra motive power or by adding balancing chains. If balancing chains are used, suitable buckets must be provided for catching the chains as the span lifts.

Equalizers, of suitable design, should be provided between the counterweights and counterweight cables. The other ends of the counterweight cables should be attached directly to the lift span.

The pitch diameter of the sheaves should be equal to at least 60 times the diameter of the cable. Clearances of at least 1/8 in. should be allowed between cables on sheaves, and sheaves should be grooved to fit the cables. Sheaves up to 14-ft pitch diameter may be made of cast steel, but larger ones should be built up of structural steel with cast steel rims and hubs. Each sheave should be fastened to its shaft by at least three keys. Sheaves at the top of the towers should be protected from the weather by hoods or housings, especially in climates where there is snow and sleet. The sheave shafts should be designed for bending, bearing, and shear stresses. The bearings should be large, properly alighned, well oiled, and not placed too far apart.

Buffers, either of the hydraulic or air types, should be provided for assisting in the stopping of vertical lift spans. Suitable buffers permit the stopping of the span with a minimum jar.

A suitable locking device should be used for the lift span to securely lock it in position before traffic is admitted to the span. The locking device must be arranged so that it can be applied or released by the operator when he is at his station. Rail locks should be used for railway and street car tracks.

Strong and substantial gates must be provided to protect the highway traffic when the lift span is raised. These gates should be in position before the lift span is raised and remain in position until the lift span is lowered and locked in place. For satisfactory operation, four gates are usually needed. The gates should be closed and opened either by the bridge operator or by special gate tenders. When gate tenders are used, small neat houses should be provided for them near each end of the lift span. Just before the gates are closed, a warning signal should be given for the benefit of the traffic.

When wide sidewalks are cantilevered outside of the trusses of the lift span, the possible effects of live loads on only one walk must be carefully considered, especially in regard to unbalanced loading and overturning moment. Proper end bearings must be provided when necessary. When a roadway or a street car track is cantilevered outside of the main trusses, provision must be made for consequent overturning moment.

The machinery for a vertical lift span is usually placed in a machinery house on top of the center of the lift span. This house should be well built with ample space for the machinery, work bench, and stove. In very large and heavy bridges, the placing of a crane in the machinery house is advisable. Suitable stairs and walks should be provided for access to the machinery house and the machinery.

In deck girder spans, the machinery may be placed below the bridge floor and between the girders, while in a half-through plate girder span the machinery may be placed below the bridge floor or outside of the girders.

When the operator does not stay in the machinery house, he should have a small house provided for him in a place where he can have an unobstructed view, in all directions, of the water and the bridge traffic. The operator's house must be large enough for the operator, operating machinery, and heater, and it should have a large amount of window space. A good stairway should be provided.

At one of the towers and on both the upstream and downstream sides, guages or indicators with large figures should be provided for the convenience of boats, showing the height of the water and also the height that the lift span is raised.

In general the operating machinery should be compactly arranged and have no more reductions than necessary. When four drums are used, one reduction should be placed at the drums. The machinery should be arranged so as to permit easy access for oiling, inspection, repairs, and replacement.

The cables used to raise and lower the lift span should be of plow steel and be composed of six strands of 19 wires each constructed on a hemp center. These cables should never be less than 3/4 in. in diameter. Preferably two cables for raising and two cables for lowering should be used at each corner of the lift span, unless the force required to move the span is so small that only one cable will suffice. These cables should be securely attached to drums in the machinery house at the center of the span and should pass from the drums over deflecting sheaves at the ends of the span and thence to the top and bottom of the towers. Some method must be provided for taking up the slack in the cables.

Whenever it is necessary to support the cables between the drums and deflecting sheaves, good rollers of not less than 6 in. in diameter should be used so that they will easily revolve and keep the cables from wearing.

For small lift spans two drums are required at the center of the span, one for the cables on each side. For larger spans, four drums are advisable. All drums should be grooved so that they may receive the cables from both ends of the span. The diameter of the drums should be about 40 times that of the operating cables, and they should be grooved so that there will be at least 1/16 in. clearance between the cables wound on the drum. The number of grooves on each drum should be such that there will never be less than one and one-half or two complete turns of any cable left on the drum.

The deflection sheaves for the operating cables should be of the same diameter as the drums and properly grooved for the cables. If there are two cables, the clearance between cables should be at least 1/4 in. A small idler sheave should be placed below each deflection sheave and toward the center of the span to prevent the up haul cable from leaving the deflection sheave when it is a little slack.

All gears should have involute machine-cut teeth, and the face width of the gear should be about two and one-half times the circular pitch. The use of bevel gears should be avoided. Worm gears may be used, provided that the gear shall have 30 or more teeth and that both worm and gear are made to run in oil. Worm gears are less efficient than spur gears.

Single-bearing frames should be used for all shafts in a unit wherever possible. Bearings should be placed close to the points of loading to eliminate bending in the shafts, and they should permit any gear to be removed without moving other gears. All bearings should be of the split type and provided with necessary shims. Four bolts should be used to bolt caps to bases except in the case of small bearings where two bolts are sufficient.

Although in general the principles of design outlined in the previous articles are applicable to lift spans recently constructed, several new ones employ a method of operation quite different from that used prior to 1935. By the use of synchronous motors coupled to the same shaft as the drive motors it is possible to place the operating machinery on housed-in decks at or near the tower tops. This arrangement also permits applying the power through suitable gear trains to the counterweight sheaves, thus eliminating the operating cables common to earlier designs.

B. Bascule

1. Mechanics

The simple trunnion principle is applied to a double leaf deck bascule. The entire weight of leaf and counterweight during the operation of opening is carried by the trunnions located approximately at the center of gravity of the mass. These main trunnions are carried in trunnion bearings which in turn are supported directly or indirectly on the masonry of the pier. Also, there is one type where in the trunnion bearings are supported on transverse trunnnon girders, which in turn are carried by the masonry of the piers. It is also common practice, however, to employ vertical posts or towers underneath the trunnion bearings, thus eliminating the necessity for a transverse girder. The use of longitudinal girders on either side of the trunnion, and parallel thereto, is also quite frequently employed. It is also possible, by detailing the counterweight with suitable recesses, to support both trunnion bearings directly upon the masonry and thus eliminate the necessity for any towers or girders whatsoever.

As the span comes to rest, the forward bearing point comes to bearing on the live load shoe, and the rear anchor lug, attached to the counterweight, engages a seat in the anchor columns, thus causing the leaf to act, under live load, as a cantilever supported at the ends.

By adjusting the shims under the live load shoe, the span may be made to come to bearing on this shoe slightly before the anchor lugs engage, thus allowing the trunnion bearings to lift slightly under live load by removing the dead load deflection from the trunnion supports.

The span is operated by means of an operating pinion rigidly connected with the pier and engaging a circular track attached to the moving leaf.

This type of structure is sturdy and simple in operation and is unquestionably one of the best types of bascule bridges in use.

C. Swing

These structures should be wisely used only for reasonably wide channels, since they require a wide central pier that would block a narrow canal. In contrast with vertical lift and many bascule bridges, the swing bridge is an indeterminate structure, the degree of indeterminacy depending upon the type of construction. Two particular types have been in common use, the rim-bearing structure and the center-bearing structure. Both require a wide central pier supporting a horizontal ring girder, on which the truss rolls as it is rotated parallel to the channel. The distinction between the rim-bearing type and center-bearing type is that a rim-bearing span is supported entirely on this ring girder, while a center-bearing span pivots on a large bronze center bearing, and is merely stabilized against overturning by contact with the ring girder. Center-bearing structures have given less trouble and are more widely used than rim-bearing structures.

IV. Inspection Considerations

A. Defects, Damage, Deterioration

Defects, damages, and deterioration typical of all steel and concrete structures can strike movable bridges, too. Therefore, most types of bridge structure defects and deterioration listed elsewhere apply to movable spans also.

Mechanical and electrical equipment include specialized areas that are beyond the scope of this manual. Since operating equipment is the heart of the movable bridge, it is recommended that expert assistance be obtained when conducting an inspection of movable spans. It should be noted that in many cases, the owners of these movable bridges follow excellent programs of inspection, maintenance, and repair.

Fatigue can be a problem with movable bridges due to the reversal of the fluctuation of stress as the span opens and closes. Any member or connection subject to such stress variations should be carefully inspected for fatigue failure.

Segmental girders and tracks on rolling lift features frequently cause maintenance problems. The repeated pressure exerted between the rolling girders and the tracks will cold work the girder or the track, causing an extension of one of the plates. Where a toothed track is used, rolling will be hampered. Where smooth plates are used, the segmental girder may lengthen, causing binding at the rear joint. Sometimes the track may flatten, warping the member to which it is connected and causing fatigue cracking.

When these bearings are out of adjustment, live load forces can shift to the main trunnion causing the overstress and increased wear.

Since it is not possible to describe all types of bascules, it should be noted that many of these bridge types transfer the full reaction from the trunnion bearings to the substructure by means of larger girders, floor beams, or heavy bracings. In such cases, the members that are exposed to these stresses should be inspected carefully for signs of distress or corrosion.

A trunnion and its bearings constitute the focal point of the leaf's rotation and, as such, are subject to deflection and wear, which can impair either the operation efficiency or safety of the bridge.

B. Inspection Items

1. Cables

Counterweight cables as well as uphaul and downhaul cables on lift or bascule spans should be inspected carefully for wear, damage, corrosion, and evidences of inadequate lubrication. To properly inspect cables, old lubricant must be removed. After inspection, the cable should be lubricated again. Check for any binding at the travel rollers and guides. Check the piers for rocking when the leaf span is lifted.

2. Counterweights and Attachments

Check the counterweights to determine if they are sound and are properly affixed to the structure. Further check temporary supports for the counterweights that are used during bridge repair.

Where steel members pass through or are embedded in the concrete, check for any corrosion of the steel member and for rust stains on the concrete. Also, look for cracks and spalls in the concrete.

Check for debris, animals, and insect nests in the counterweight blocks. Where cable counterweights or balance chains are employed, check links, slides, housings, and storage areas for deterioration, for adequacy of lubrication, where applicable, and for protection. Determine whether the bridge is balanced and whether extra weight blocks are available. A variation in the power demands on the motor, according to the span's position, is an indicator of balance problems. Paint must be periodically removed from the lift span proper; otherwise, the counterweights will eventually become inadequate.

3. Drainage

Check to determine whether the counterweight is properly drained. On vertical lift bridges be sure that the sheaves and their supports are well-drained. Examine any portion of the bridge where water can collect.

4. Piers

Check the piers for rocking when the leaf is lifted. This could be an indicator of a serious problem and should be reported at once.

5. Warning Devices

Check the operation of safety gates, barriers, and warning signals. Be sure they function properly and that the warning signals give sufficient notice to permit vehicles to clear the automatic gates. Determine whether the safety gates and barriers are sound and well-maintained. Note any decayed areas at bolt and other connections. Note whether either needs to be replaced or repaired. Note the locations of the safety gates in relation to the warning light, signs, and bells, and the bridge opening itself. Check the warning lights with regard to location to determine whether they can be easily seen by motorists.

6. Machinery

On all movable structures the machinery is so important that considerable time should be devoted to its inspection. The items covered and termed as machinery include all motors, gears, tracks, shafts, linkages, overspeed control, brakes, and any other integral part that transmits the necessary power to operate the movable portion of the bridge. The inspection of the next ten items and items similar to them should be made by a machinery or movable-bridge specialist.

Check the alignment of all gears, locks, and other interlocking mechanisms. Check the adequacy of the lubrication of all movable parts, particularly where meshing or contact occurs between the movable parts. Also check the schedule of lubrication to determine whether the frequency of lubrication is sufficient. On live load bearings, check the wedge (lock) linkages for loose knee pins and for excessive play. Note the closing and releasing of the wedge locks or pin locks for proper functioning. Check all gears for cracks including the teeth, spokes and the hub. Inspect all shafts for twisting, strain, and for play within bearings. Check the keyways on the shafts and the gears for looseness. Check the keys for looseness also. Check the braces, bearings, and all the housings for cracks, especially at welded joints. Inspect the concrete for cracks in areas where machinery bearing plates of braces are attached. Note the tightness of bolts and the tightness of other fastening devices used. Check all brake devices for proper functioning. Check to see whether stops are used and are needed.

7. Motors and Engines

The inspection of motors and engines should be made by mechanical specialists. The inspection should include the check of the following conditions. If a belt drive is used, check for wear and slippage. Note the condition of all belts and the need for replacement, if any. If a friction drive is used, check for wear and for uneven bearing areas. If a direct drive is used, check all bracings and bearings for tightness. If a liquid coupling is used, check the fluid to ascertain that the proper quantity is used. Look for leaks.

C. Swing Bridge Inspection Items

Check the wedges and the outer bearings at the rest piers for maladjustment. This can be recognized by excessive vibration of span or uplift when load comes upon the other span. Check the live load wedges and bearings located under the trusses or girders at the pivot pier for proper fit. Check the teeth of all gears for wear and for proper alignment. The meshing of gears should be checked so as to determine whether gears are climbing one upon the other.

On center-bearing swings, check the center pivot, the housing, the tracks, and balance wheels for fit, wear, pitting, and cracking. Check for proper and adequate lubrication.

On rim bearing bridges inspect the center pivots, the rollers and the roller shafts, and the guide rings or tracks for proper fit, wear, pitting, and cracking. Check for proper and adequate lubrication.

D. Bascule Bridge Inspection Items

Check the center locks on double-leafed spans, and note whether there is excessive deflection of the center joint or vibration on the bridge. Inspect the locks for fit and for movement of the leaf (or leaves). Check lubrication and loose bolts. Check the lock housing and its braces for noticeable movement. This can be done by observing the paint adjacent to it for signs of paint loss or wear. Check the differential vertical movement at the joint between the two leaves under the passage of heavy loads. Check the joint between the two leaves for adequate clearance. Check the front live load bearings to determine whether they fit snugly. Also observe the fit of tail locks at rear arm and of support at outer end of single leaf bascule bridges.

Check the bumper blocks and the attaching bolts for cracks at the concrete bases. Check the counterweight well for excessive water. Check the condition of the sump pump, the concrete for cracks, and the entire area for debris. Check the brakes, limit switches, and all stops for excessive wear and slip movement. Note whether the cushion cylinder ram sticks or inserts too easily. Check the shaft or trunnion bearings for excessive wear, lateral slip, and loose bolts. Check shear locks for wear. Excessive movement should be reported and investigated further.

On rolling lifts, check segmental rim and girder, and the track plate and girder for the following. Extension of the rim plates and track plates in either direction. Wear and poor fit of toothed rim and track plates, including cracking in the corners of the slots, in the rim plates, and fractured teeth. Distortion of rim and track plates including curling of edges and separation of plates from their supporting girders. Cracking at the fillets of the angles forming the flanges of the segmental and track girders. Cracking of the concrete under the track. Looseness between walking pinion gear and top rack. Check top rack for lateral movement when bridge is in motion. On heel trunnion (Strauss) bascules, check the strut connecting the counterweight trunnion to the counterweight for fatigue cracks. On several bridges, cracking has been noted in the web and lower flanges near the gusset connection at the end nearer the counterweights. The crack would be most noticeable when the span is opened. The rack and pinion should be inspected for gear wear, cleanliness, and corrosion.

E. Vertical Lift Bridge Inspection Items

Check span and counterweight guides for proper fit and free movements. Span guides are usually castings attached to suspended span chords which engage a T section attached to the tower. Counterweight guides are angles or tees attached to the tower and engaging grooved castings attached to the counterweights. These grooved castings must be inspected closely for wear in the grooves. Check cable hold-downs, turnbuckles, cleats, guides, clamps, and splay-castings. Check the motor mounting brackets to ensure secure mounting. Check alignment and wear of cables, drums, and sheaves. Note whether cable is running properly in sheaf grooves. Recommend the replacement of all grayed or worn cables. Look for any obstructions to proper movement of cables through pulleys, etc. Check spring tension, brackets, braces, and connectors of power-cable reels. Check travel rollers and guides, brakes, limit switches, and stops. Since the machinery room is usually under the main deck, check the ceiling of the machinery room for leaks or areas that allow debris and rust to fall on the machinery. Survey lift bridges, including towers, to check both horizontal and vertical displacement. This should identify any foundation movements that have occurred.

F. Control House

Consult with the bridge operator to ascertain whether there have been any unusual developments in the operation of the bridge. Note where the control panel is located in relation to roadway and waterway. Note whether the bridge tender has a good line-of-sight view of approaching boats and vehicles. Note whether structure shows cracks. Determine whether it is windproof and insulated. In some cases only control boxes are provided, without a bridge tender. Note this situation and check the security system. If controls are separate, note description of bridge tender's house or shed, and include its condition as well as the information about the control house. Note whether alternate warning devices such as bull horns, lanterns, flasher lights, or flags are available. Note whether all Coast Guard, Corps of Engineers, and local instructional bulletins are posted. Check for obvious hazardous operating conditions involving the safety of the operatory and maintenance personnel. Check for any accumulations which may be readily combustible.

Check controls and electric panels on movable structures. An electrical specialist should be available for this part of the inspection. Check controls while bridge is opening and closing. Look for excess play and sparks. Check electrical cabinets for loose wires, heaters, and bunched-up wires. Note debris or material hidden in cabinets. Inspect the electrical system of the bridge including the wiring, conduits, motors, and lights. Check for worn or broken lines. Check for any existing hazardous condition. Check for rusted-out or mismatched members. Determine whether the controller is outdated or parts need replacement. Determine whether electrical interlock is working. Check whether panel doors are secured. Note whether bridge tender has any complaints about the panel. Check span speed-control resistor banks for overheating.

G. Main and Submarine Cables

On the main cables note the condition of the power lines coming to the bridge. Where high voltage lines come all the way to a transformer in the control house, check that the main lines or cables are fully insulated and out of reach of the public. Where transformers are on a power pole near the bridge, check the rigidity of the pole, guy lines, ground line, and cable to bridge. Check the transformer in the control house, if any, for bracing, high voltage cables insulation loss, leaks, and cable protection. If a cable is attached to the bridge, check anchors, clips, concrete bases, insulators, or armor, for attached growth or debris. Determine whether the line or transformers should be replaced or relocated. Check lightning arrester device for signs of distress.

Submarine cables should be labelled for size and number of conductors within the submarine cable. Indicate the number of conductors being used, the number of spares that are available, and the number of conductors that have failed. This should equal the total number of conductors in the cables. Note whether the cable is protected from boats and public, and whether it is behind the fender system. Note whether the cable is kinked, hooked, or exposed either above or below the water. Note whether the ends are conditioned and protected from moisture. Check the cable at tidal areas for excess marine or plant growth. If cable has been spliced, note conditions of box seal. Inspect clamps and securing clips.

H. Auxiliary Power

Operate auxiliary power or crank and note condition and reliability. On double-leaf bascules, note whether both sides have auxiliary power systems. On hand-cranked systems: determine whether standing platforms are free of grease and debris; determine the number of men needed in this operation; determine whether a portable generator powered mechanical device can replace the manpower needed to operate the bridge.

Chapter 9

SERVICES

I. Signing

It is the responsibility of the bridge inspector to check all signing associated with the bridge. That is, signing on the roadway, waterway, or that which is affixed to the bridge itself. Therefore, he must be familiar with the types of signing that are associated with bridges. The following discussion, reviews the types of signs that may be encountered during on-site bridge inspection, their probable locations and their effectiveness in conveying a message.

Bridge signing performs interrelated functions. Most importantly, signing warns of predetermined hazardous conditions for traffic in regard to the bridge structure. Bridge signing serves to promote the safety of the public and the bridge as well. It informs users of widths, overhead clearance, load limit, bridge conditions, etc. Other functions and kinds of information that are provided by signs are highway routing, directions, destinations, and public convenience notices.

The bridge inspector will be primarily concerned with the presence and condition of bridge signing for the purpose of maintaining or increasing the safety level. These signs may be required by federal, state, or local law. Typical of such signs are those which inform the public of such things as: weight limit, vertical clearance, lateral clearance, and speed. Signs which inform the public as to route, directions, destination, public conveniences, or points of interest, if present, should concern an inspector only from the standpoint of their physical presence and their condition and effect on the bridge.

A. Classification of Signs

Signs may be classified by the function they serve:

Black and white regulatory signs provide for the public a notice of traffic laws or regulations that pertain on the bridge structure. Rectangular is normally the shape.

Black and yellow warning signs call attention to conditions on or adjacent to the bridge that are potentially hazardous to the public. Diamond is the normal shape.

Guide signs provide information that is nice to know, such as routes, directions, etc. These signs may be black on white or white on green or blue, and are normally rectangular in shape.

Black and orange signs indicate construction or other temporary conditions.

Bridge markings are normally one of four general types. End markers on a bridge usually are a reflectorized diagonal black and white striped hazard board on a pole or reflectorized paint on the ends of the structure. Reflector type markers may be used at bridge ends, piers, and abutments, or as delineators. Delineators are glass or plastic reflectors usually yellow, which may be found mounted vertically alongside the roadway to help guide traffic. Painted pavement marking such as lines, curb lines, or no passing lines are normally reflectorized. Reflectorized and non-reflectorized formed-in-place wet pavement markers are becoming more common.

B. Location of Signs

Each state has its own regulations concerning the placement of signs. The "as built" plans for each bridge should provide the necessary information on which to base the inspection of bridge signing. However, there are some general considerations that should be understood about the location of signing which should improve the inspection coverage of signing.

A weight-limit sign should be located on the shoulder just ahead of the bridge. It is a regulatory sign. Notice of weight-limit should also be posted at the last intersection.

The vertical-clearance sign is usually painted on the bridge superstructure or on a sign attached to the bridge superstructure. In the case of bridges over navigable channels, the sign will most likely be attached to the far end of a dolphin and fender system. It should be a sufficient distance ahead of the structure on the side of the roadway or channel to allow the driver or mariner to react.

In rural areas, an additional sign pertaining to any restrictions of the bridge should be located before an appropriate intersection to permit vehicles, if necessary, to conveniently avoid the bridge.

The lateral-clearance sign indicates to the user such warning information as narrow bridge or one lane bridge. The need for one lane bridge signing should be obvious. Narrow bridge signs should be used whenever the bridge width is less than that of the approach roadway. These signs will normally be located on the roadside just ahead of the bridge.

The placement of signs will depend on the conditions of the area or site in which they are located. In general, sign faces should be positioned so that reflected light does not impair the vision of vehicle drivers.

In general the signs you may encounter will have their bottom edge 5 ft from the pavement or higher. Markers, of course, may be at any height from the ground level to the top of a structure. When located at the side of the approach roadway, the vertical edge of the sign nearest the roadway should be at least 2 ft from the prepared surface including the emergency area.

In general sign supports will range from simple timber or steel posts to two or more steel I beams, to sign bridges from which signs are hung and lighted. In addition some signing is attached to the structure directly.

C. Inspection of Bridge Signing and Marking

When inspecting bridge signing and marking, the following items should be checked.

Presence

Legibility: It is capable of being read. Clean and unspoiled by vandals.

Visibility: Can be seen. Not hidden by vegetation.

Need: Signs are ordinarily not needed to confirm the rules of the road.

Condition: No deterioration and protected by guard railing from vehicle impact.

Proper mounting: No missing elements which make the sign or marking incomplete or hazardous.

Proper support: No collision damage, or deterioration from corrosion, insect attach or fungus decay.

II. Lighting

This topic will provide information concerning the various lights that may be encountered and what the inspector may expect to see. Most people know lighting exists on bridges, but they often take it for granted. That is, we assume there will be lighting adequate to illumine the bridge, the signs on thc bridge, and in some cases, to assist in the control of traffic. Few people have noticed red lights on the top of a bridge or the green lights and red lights that are on the sides of some bridges. Most lights that may be observed to be out on a particular night will most probably be lit the next night, not because a bridge inspector reported the outage, but more likely because a policeman, state trooper, state highway department employee, or a passerby reported the outage.

A. Types of Lighting

There are basically five types of lighting that may be encountered during bridge inspection: Highway or "whiteway", aerial obstruction, traffic control, navigational, and sign lighting.

Highway or whiteway lighting presents few problems. In most cases, the lighting has been specified by the State Highway Department and installed and maintained by the state or local electric utility company. The fixtures are similar to those found in large parking areas and along city streets and highways. Most of the highway lighting standards will be made of steel or aluminum. In some cases, concrete or timber shafts and bases may be encountered.

At the top of the light standard is the lamp or luminaire. This, of course, is the key to effective lighting. The purpose of bridge whiteway lighting is to provide for maximum roadway safety at night. Anything which impairs the effectiveness of the luminaire will impair the purpose of lighting. A dirty glass reduces the effective coverage of the light.

The shaft will be attached at the bottom to an achor base. In the case of steel and aluminum shafts, the shaft is fitted inside and welded to the base. In the case of concrete, the base is normally cast as an integral part of the shaft. Where the standard is exposed to vehicular traffic, one may find a frangible or breakaway type base and/or barrier-guard railing.

The purpose of aerial obstruction lights is to alert pilots of aircraft that a hazard to their aircraft exists below and around the light. The light is red, is visible all around and from above, and should be operating during periods of darkness and of poor visibility.

These lights are located on the topmost portion of any bridge considered by the Federal Aviation Administration to present a hazard to aircraft. Depending upon the size of the bridge, there may be more than one light.

Traffic control lighting is a necessity for a movable bridge, but may also be found on fixed spans. Checking these lights is the responsibility of the bridge tender, but the bridge inspector provides a cross check. Traffic signals to control traffic, include signals such as lights, bells or horns, flags or a combination of any or all of these. They must be inspected to ensure they are visible or audible and operating properly. The bridge may be equipped with warning gates; lights along the gates must also be checked.

The U. S. Coast Guard determines requirements for the type, number, and placement of navigational lights on certain bridges. The purpose of navigation lights is for the safety and control of waterway traffic.

There are two possible situations posed by fixed bridges, the first is a narrow waterway with no obstructions, while the other is a wider waterway with obstructions in the form of piers. Both have red and green lights. The red lights create an arc of visibility of 180° to an approaching mariner, while the green has all-around or 360° visibility. In the case of the narrow waterway that has no obstructions, the lights are attached to the superstructure on both sides to tell the mariner how best to navigate past the bridge.

In the other situation, the lights serve the same purpose but now the red lights are located on the pier.

The through swing bridge has specific lighting requirements. When closed, the swing span shows all red beams of light to the approaching mariner. When open, it shows green or green over red. In the closed position the two green lights show red to the mariner. The lights are fixed in position on the span and the colors separated by an internal light separator. Damage to this light could cause a serious accident.

B. Inspection of Bridge Highway Lighting

The state is obligated to the public to provide roadways that are safe at night. The responsibility within a state for providing lighting will vary. The responsibility may be assigned to the state highway department, the public works department, a county engineer, or other local official. In any case, authority is normally given, along with responsibility, to contract for the installation and service, including maintenance, of lighting. In most cases then, some public utility organization is responsible for inspecting and maintaining highway lighting.

A bridge inspector should look for the following when checking the bridge lighting:

Broken and/or dirty luminaires: Significant accumulation of dirt reduces output.

Corroded standards in the case of steel and aluminum.

General aesthetic appearance: Need for paint.

Spalled concrete at base.

Vehicle-collision damage.

Integrity of standard: Luminaire appears to point in proper direction for illumination of roadway, and base is securely bolted to the foundation. No damage caused by excessive vibration.

Hand hold covers secure.

Clogged weep or drain holes.

Corrosion of and near the light.

Broken glass.

Operation: Turn on the lights or have them turned on to be assured that all work.

Exposed wiring: Check for deterioration, but do not touch a wire that appears to be defective. Call the supervisor, and have a qualified electrician determine the condition and take immediate corrective action.

For bridges over navigable channels, check to see that the required navigational lights are properly installed in their intended position and are functioning. The U.S. Coast Guard requires that these lights be turned on during periods of low visibility and darkness. Consequently, in some cases they remain lit 24 hr a day or switch on in response to a photoelectric cell, or they are tied into an automatic system with other bridge lighting. In any case, make arrangements to have the lights turned on for the inspection.

III. Utilities

Utilities consist of such public services as gas, electricity, and water. Utilities must be transported by means such as a pipe, conduit, or power cable. These transport mediums are normally buried in the ground, strung along poles or placed on supports above ground. Most state highway departments have a policy against carrying utilities on their bridges or through their drainage structures.

However, as often is the case, waivers are sometimes granted. Therefore, utilities are often referred to by the term encroachments.

The inspector probably will encounter some type of utility on quite a number of the bridges. These utilities should be recorded in the report so that they may be cross-checked to see if permission was granted for the use of the bridge. Utility installations through drainage structures are not normally permitted. Again, however, the inspector may very well encounter utility installations during inspection of drainage structures. These should be recorded.

However, a highway department may not desire utilities on its bridges or through its drainage structures. The placement of apparatus for utilities may overload the bridge. In addition, indiscriminate installation of utilities makes the maintenance of the bridge more difficult. Finally, some flammable materials, such as natural gas, if permitted on a bridge and installed improperly, constitute a potential hazard to safety and to the integrity of the structure. In the case of drainage structures, without some controls, indiscriminant placement of utilities could interfere with the primary function of the structure, that is, its ability to provide proper drainage. Therefore, it is a good policy not to allow public utility installations to be placed on or attached to bridges or placed through drainage structures. As stated previously, exceptions to this policy are made, but those state highway departments allowing some leeway are able to maintain careful control over the number, types, and installation of utilities that become exceptions to their policy. This is because the department is notified of the exception and has a procedure for approval, and approvals should be with the bridge records. Utilities placed upon or attached to a bridge or through a drainage structure without permission are considered encroachments.

Normally, a utility is placed upon a bridge in a manner specified by the State Highway Department. In some cases, the bridge design accommodates the installation of utilities. The inspector will find this the case where pipes are led through abutments or retaining walls. Thus, when it is known ahead of time that certain utilities will be carried by the bridge, this fact is taken into account in the design. In these cases, well-supported utilities are found. Most of the problems are encountered where utilities have been added after bridge construction. For example, it is often easier to attach a power cable on the side of the superstructure than to place it more safely beneath the superstructure and out of sight.

The owner of the utility is responsible for its inspection and maintenance. Some larger utility companies are faithful in this regard, but some smaller companies are remiss in their duties. In either case, the inspector must check utilities to protect the public interest in bridge safety.

In preparation for the inspection of a bridge, the inspector should review the available records and reports pertaining to the bridge. Make note of those utilities or, as they may be called in some states, encroachments that may be encountered on or near the bridge. There may be a separate page or sheet with information regarding utilities. In most cases, it will include information on utilities or encroachments, in the most immediate area, though not fastened to the bridge, such as a sewer line crossing the right of way and buried in the channel beneath the bridge. Often, photographs of the utilities are available. With this information, the inspector should be better able to recognize encroachments or unauthorized installations. During the preparation review, the inspector should:

Note the location of each authorized utility.

Determine how and when to inspect the utilities. For example: Is a snooper required, or can the utilities be checked from the available catwalk?

Will it be necessary to crawl along the pipes to properly inspect the utility and its supports, or can they be checked from a catwalk or snooper?

The answers to these questions should be determined in conjunction with the best way to properly reach and inspect the caps, bearings, floor system, etc.

The following items should be reviewed during inspection:

Unauthorized installation.

Pipe, ducts, etc: leaks, breaks, cracks and deteriorating coverings.

Supports: signs of corrosion, damage, loose connections, general lack of rigidity, need for padding. Rigidity or need for padding can be checked during the passage of vehicular traffic. Check whether vibration or expansion movements are causing cracking in the support members.

Abutments: space between pipe and sleeve, or pipe and sealed area where utilities pass through abutments for leaks.

Water or sewer pipes: leaks from water or sewer pipes located above the deck or on top of beams may be such as to cause serious corrosion of the deck or beams. In addition, the area beneath these pipes should be checked for damage.

Separation of flammables: Ensure pipes transporting flammable products are isolated from electric utilities. Report breakdown in protective systems of either and in addition, report the need for relocation of one or the other.

Clearance: Ensure utilities located beneath overpass type bridges provide roadway clearance and are properly supported to prevent danger to traffic passing under the bridge.

Waterway obstruction: Report any utility of encroachment obstructing the waterway or positioned so as to hinder the removal of drift material during periods of high water.

Pressurized pipe: Check for leaks in the encasement material, drains, vents, and shut-off valves, if present.

Electrical conduit wiring: check condition of conduit for rust, missing sections, weakened supports, deteriorated shielding, and insulation on any power cable. Do not touch bare wire.

Ease of maintenance: Report any adverse effect on bridge maintenance or operations presented by the utility.

Impairment of structural integrity: Check and report features of the utility and its support that impair the structural integrity of the bridge.

Paint: Has the installation of a utility caused damage to the bridge paint system?

Aesthetic appearance: Provide a general comment in your report pertaining to how the utilities affect the design or artistic appearance of the bridge.

Chapter 10

INSPECTION EQUIPMENT

I. Safety

The federal bridge maintenance inspection program was established to ensure that the bridges on the public transportation systems provide safe means of travel. Likewise, the individual inspectors must maintain a program of safety for themselves and fellow bridge inspectors while on the job. Occasionally, individuals do not observe safe working practices because of an attitude that it's a sign of weak character not to simply accept challenges and danger. Some fail to observe safety through carelessness. Individuals with such traits must be quickly educated or removed from the bridge inspection crew. The safety of each individual depends upon his own actions and the actions of his fellow crew members.

An individual may have an accident when working even under the safest possible conditions, especially if he does not assume responsibility for his own safety. However, a person working under less than ideal safety conditions can usually avoid accidents if he will assume responsibility for his own safety. Whatever the conditions, the individual must consider that it is to his advantage to work safely. If he is not willing to accept this responsibility, the chances are that he would pay little attention to safe practices and procedures. Consequently, it can be expected that such an individual is more likely to have an accident.

Consider when accidents occur. Any task that is repeated many times becomes habitual, and motions become automatic. Most accidents happen during the performance of these routine tasks, rather than during the performance of an unfamiliar task.

The consequences of an improper attitude toward safety are too great to ignore or fail to give the fullest consideration. Consider the consequences:

Death: Loss of the individual's life plus hardship for the family that is left.

Permanent total disability: The family must not only provide for itself, but must take care of the disabled individual.

Permanent partial disability: The individual must bear considerable expense and often must readjust his life and find some new work to support his family.

Other injuries may be only temporary but each injury costs the individual and his employer time and money, often valuable equipment is lost, and the productive time of the injured individual is lost.

A. Causes of Accidents

Almost all accidents are either directly or indirectly attributable to human failings. Man is not a machine; his performance is not fully predictable, and he often times makes mistakes. Anyone can make a mistake—the designer of a building, the builder of a bridge, and most probably, the individual who has the accident. An accident may be partly due to worry, grief, ill health, bad temper, frustration, intoxication, or other physical and mental states. Set standards from which you will not deviate.

1. Safety Precautions

Inspecting a bridge can be very dangerous if the proper safety precautions are not observed. Some considerations include the following:

A bridge inspector must be as alert as possible at all times. Alcoholic beverages or narcotics before or during work cannot be condoned. Intoxicants impair judgment, reflexes, and coordination.

Electricity is a potential killer. Assume, until tests prove differently, that all conductors or electricity are hot (alive). The conditions encountered on many bridges (metal materials of construction, humid atmospheres) are conductive to electric shock.

Try to work in pairs. An inspector should not take any action that may leave him without help in case of an accident.

A safety skiff fitted with life preservers and life lines should be provided when working over bodies of water from which rescue by other means would be too time consuming, or impossible.

Use a flashlight to illuminate dark areas prior to entering, as a precaution against snakebite and stinging insects. First aid kits should be available.

2. Proper Personal Equipment

The inspector should never forget his own personal safety equipment. The proper equipment and proper use of that equipment is necessary for personal and team safety. Personal equipment should include the following:

Hard hat with chin strap, to protect the head from falling objects or bumps from accidental head contact with bridge members.

Eye protection when working in an eye-hazardous area, to protect the eye against injury from flying debris.

Life jacket when working over water, to protect the inspector in case he falls into the water. Even a good swimmer may be unable to save himself, because of the effect of the fall.

Safety line and belt when working at heights over 20 ft, above water or traffic or at other times you consider appropriate, to protect the inspector from being seriously hurt in the event of a fall. Safety belts are awkward but should be worn when practicable with a safety line attached securely to a member capable of supporting the inspector if he should fall. An inspector may find that a safety belt is impractical under certain conditions; however, it is still his decision as to whether or not he will wear one.

Safety shoes with a "steel toe" protect the toes from falling objects. They should be of the high top type so as to give additional ankle support when climbing. Non-slip soles prevent the inspector from slipping only if the soles are kept free of grease and oil.

Gloves and warm clothing when working in cold weather protect the hands from the cold and from cold bridge material. Warm clothing allows the inspector to complete his inspection without being preoccupied with the cold. While this clothing should provide warmth, it should not hinder the inspector's freedom of movement.

Cold weather brings on other problems such as the difficulties of completing reports. Tape recorders are of some help; however, they too may be a hindrance in high or otherwise difficult areas.

B. Traffic Safety

The inspector must take thorough precautions when working in traffic areas. The inspector has no control over the driver of the vehicles and must assume the driver may not always be alert or careful. Therefore, the inspector should be alert and maintain strict safety practices.

Make sure proper signs are posted. The inspector must warn the driving public that all lanes of traffic will not be in use, primarily for the safety of the inspector but secondarily for the safety of the motorist. Florescent orange flags should be affixed to each warning sign used.

Try to work during slack periods. The inspector will have less difficulty in completing his inspection, plus the inspector's relationship with the public will be improved.

Make sure that traffic lanes are closed "gradually" with enough room so that cars can merge with no problems. Under relatively normal conditions of speeds and volume, and where adequate advance warning of a lane obstruction has been provided, a taper rate of 1:20 should be sufficient to permit traffic to shift safely from one lane to another. Where speeds or volume are high, this rate should be substantially decreased, to about 1:40; expressways may require even longer tapers. Where traffic is stopped or considerably slowed in advance of the transition, as by flagman, and where the lane change does not involve a merging of traffic streams, the taper may be very short, just long enough for traffic to turn comfortably into the appropriate lane. Cones, drums, or barricades are used to funnel traffic into the appropriate lane. When a lane of traffic or portion of roadway is blocked by equipment, the following safety measures should be taken:

Plan work to reduce unsafe working conditions and actions.

Post advance warning signs with fluorescent flags affixed.

Allocate high-visibility vests to employees working on or near roadway.

Use buffer vehicles on each major traffic approach to work.

Use traffic cones to channel traffic a safe distance from work forces.

When necessary, use flagmen to establish positive traffic control in vicinity of site.

Use revolving lights and flashing lights on buffer vehicles.

Keep work force within protected area.

Inform local traffic control authority prior to disrupting traffic flow.

C. Climbing Safety

Check all equipment before you start to climb. If you normally wear glasses, with the exception of bifocals, you should wear them when climbing. Bifocals should not be worn because of possible split vision when looking down. The inspector who wears bifocals should acquire a separate pair of single lens glasses.

Know what you are going to do on the bridge before you start to climb. If the inspector has no idea where he is going or what he should do, then he stands a chance of finding himself in a position from which either he may fall or he may be unable to get down.

Go about your job in an orderly manner. If the inspector is organized, he will not feel a need to take dangerous short-cuts in order to complete the inspection on time.

Scaffolding should be of ample strength and inspected daily for cracked or weakened areas. If the inspector accepts inadequate scaffolding or if he does not inspect it daily, then he stands the chance of being injured in a fall.

Ladders should be used only when necessary, and when used they should be blocked at the bottom. Before each use they should be inspected for cracked or weak areas. Ladders are very dangerous and cause many accidents yearly. Two of the biggest problems with ladders are rungs that break when least expected and bottoms that slip so that the ladders fall. Ladders should not be used as horizontal bridges, nor should they be used to reach objects that are clearly too high.

Planks should never be used singly. If one plank is used and it breaks, then the person will fall. However, if two or more are used and one breaks then you step to the other plank. Plank ends should be securely attached to whatever they are resting upon. The inspection of a plank is similar to that of a ladder and should be made before it is used.

Keep working area clear of dirt, tools, and other extraneous objects. If the area is not clear then the inspector stands the possibility of tripping on an object. In addition, something may fall from the inspection work area and strike someone below.

Do not climb when emotionally upset. The inspector who climbs must have complete control of himself, otherwise he increases his changes of falling.

If possible, work from a traveler, catwalk, or platform truck. Not only are these methods faster but they are also much safer. Do not forget, however, to check that the traveler, the catwalk, or the platform truck are in a safe operating condition. Do not forget the possibility of deterioration of hand-hold rods and cables that are attached to the work platform.

II. Tools and Equipment

A. Typical Inspection Equipment

The bridge inspector should not only be familiar with the different types of tools and how to employ them, but with the capabilities of each as well. The first tool to be considered will be a chipping hammer. The chipping hammer is used for a variety of things. The soundness of timber may be determined by tapping the timber and listening for a "hollow" sound. Timber should be checked very carefully above and below water lines and in other areas where moisture is likely. Paint or corrosion may be chipped away so that suspected deterioration of the base material may be investigated. Concrete may be checked for hollow areas by tapping concrete and listening for a hollow sound.

Paint may be removed with a scraper so that areas of suspected deterioration of the base material may be investigated closely. The paint is scraped away by placing the flat side of the scraper against the paint and moving the scraper back and forth using suitable force until the paint is removed. A scraper may also be used to scrape away dirt and debris from areas that cannot be cleaned by other means such as wire brush or whisk broom. Another use of the scraper is for removing marine growth from around the water line of a pile.

A wire brush can be used to remove corrosion from areas that cannot be adequately and quickly cleaned with other tools such as a paint scraper. It may also be used to clean debris from the area to be inspected so that the basic material will be exposed. The wire brush is moved back and forth much the same as a stiff brush.

A pocket knife can be used to check timber for soft spots after a suspected area has been found by tapping with a hammer. This check would be done by probing the wood with the knife blade to see if there are unsound areas. Be extremely careful that the knife blade does not close on your finger. Holes made in the protective covering of the timber being inspected should be treated with a preservative to prevent the wood from being attacked. A knife can be used like a paint scraper to remove paint, corrosion or debris from confined areas.

The long, thin, tapered ice pick makes an excellent tool for testing timber for soft spots. Again, as with the pocket knife, the ice pick should probably be used after a hammer has detected hollow-sounding areas. If extensive damage is suspected, then take a test boring with an increment borer. Prior to starting the test boring, make a hole in the timber with a spike or an ice pick so the increment borer will start easily. Holes made in protective covering of the timber should be protected with a preservative.

An increment borer is used when deterioration is suspected below the surface of the timber. Borings should be taken very selectively so as not to further weaken the already damaged timber. If all indications are that the wood is sound, then test borings should not be taken. Once the boring has been made, save the boring sample and make sure that creosote plugs are inserted into the hole made by the borer.

A flashlight is extremely helpful when used with an inspection mirror. The flashlight beam is directed into the area to be inspected. The mirror and the flashlight are maneuvered until the area to be inspected is clearly visible.

A plumb bob with twine can be used to determine vertical alignment. To use, secure the plumb bob string to the upper point of the item being checked. When the plumb stops swinging, measure the difference between the plumb bob string and the member to determine variance from the vertical. Although a plumb bob can be used to get a close approximation of the vertical alignment of bridge components, survey equipment is required to check absolute alignment. Plub bob and twine can also be used as a reference point from which to measure deflection of a member such as that resulting from impact.

A level can be used to give a rough indication of vertical and horizontal alignment. Again, as with the plumb bob, absolute measurement of alignment is normally accomplished using survey equipment. The level, like the plumb bob, makes a good reference point from which to measure the extent of damage or deterioration. The level can be used as a horizontal reference or it can be used as a straight edge when a vertical or a horizontal reference is not required.

A defect such as a crack can be measured for length, width, and depth, with a folding ruler. To measure section loss, determine the average depth of a defect and multiply it by the average width of the defect. Then compare the amount of material lost with the original cross section area to arrive at a percentage of loss. If the determination of section loss is to be made on a circular pile for example, the following method may be used. Measure the diameter of the pile at a location that will give the original diameter. If the diameter is 12 in. then the radius is 6 in. and the area would be $a = \pi r^2$ or 3.14 x 36 which would equal 113 in.^{2} Measure the diameter of the pile at the area where the most damage has taken place. If the diameter is 8 in. then the radius of 4 in.; then, the area would be 3.14 x 16 which equals 50 in.^{2} The amount lost would be 113 minus 50, or 63 in.^{2} The percentage of section loss would be 56%

A pocket tape can be used for measuring purposes essentially in the same manner as the folding tape. If a determination of section loss is to be made with a pocket tape, place the tape around the pile to measure the circumference of the original section of pile. If the circumference equals 37 1/2 in., then

$$r = \frac{c}{2\pi}$$

$$\text{Therefore, } r = \frac{37\ 1/2}{2 \times 3.14} \text{ or about 6 in.}$$

$$A = 3.14 \times 36 = 113 \text{ in.}^2$$

If the circumference around the poorest section is 31 1/2 in., then

$$r = \frac{31.5}{2 \times 3.14} = 5 \text{ in.}$$

$$A = 3.14 \times 5^2 = 78.5 \text{ in.}^2$$

$$113 - 78.5 = 32.5 \text{ in.}^2$$

The percentage of section loss is equal to 29%.

A longer tape such as a 100-ft tape is needed to measure lengthy bridge elements.

A larger screwdriver is used for items such as raising an expansion plate cover so that the expansion joint may be inspected. It may also be used to check weep holes for clogging.

A sounding line or long indexed pole (measured in feet) is often needed. Sounding line may be used to measure the depth of the water so that areas of scour may be detected. The sounding line is dropped into the water and reeled out until the bottom is struck. The depth of the water will be determined by noting the index mark on the line at the water level. Sounding line is very inaccurate when used in areas of swift current because the current will deflect the line much as a leaded fishing line will be deflected when it is pulled by a fishing boat. A long marked pole is used much as a sounding line; however, it is only usable in shallow water.

Thermometers are used to measure the temperature needed for recording the deflection of bearings and expansion joints. Contact thermometers are used to measure the temperature of specific areas on the bridge as it may differ from the air temperature. Contact thermometers are used by placing the base of the thermometer against the material for which temperature is to be determined.

Dialed calipers can be used to determine the amount lost due to deterioration. Place the calipers around a good section of material and take a reading. As an example, the measurement may be 1 in. Then place the calipers around the section which has deteriorated and take a reading which for example measures 3/4 of an inch. The thickness loss at that point would then be 1/4 in. or 25% of the original cross section. Calipers are also very useful for measuring the deterioration along the flange of a steel beam or girder.

Dye penetrant is used for crack detection. The principles involved in dye penetrant inspection are simple, and easily applied. First, the area is cleaned. Next, a bright, red-colored liquid penetrant (like penetrating oil) is applied to the area to be inspected. About 20 to 30 min is allowed for the penetrant to seep into any cracks or other defects which may be present. Then the excess penetrant is removed and a white-colored developer applied over the area. The developer acts as a blotter, drawing out a portion of the penetrant which has seeped into the defect, causing a bright red outline of the defect to appear in the white developer. Dye penetrant inspection is limited to identifying the location and extent of surface cracks and surface defects. Dye penetrant is not suitable for determining the location of subsurface defects. Nevertheless, it is an excellent tool to assist the bridge inspector in evaluating the presence or extent of small and/or hairline, fatigue cracks suspected in bridge elements, welds, or connections.

Commercial dye-penetrant kits are readily available and inexpensive. Most kits contain a cleaner, penetrant, and developer. Each is provided in a spraytype can, and all may be purchased in a hand-carried kit. The inspector need supply only wiping cloths to remove the excess penetrant from the area being inspected.

B. Access Equipment

A hydraulic lift is one of the more versatile pieces of equipment used in bridge inspection. Advantages of hydraulic lifts include the following:

Mobility: The hydraulic lift has a particular advantage in that it is vehicle mounted and, therefore, can move from bridge site to bridge site with relative ease and quickness.

Increased range of movement. The inspector's platform may be moved underneath the bridge deck and may be positioned so as to allow for inspection of the superstructure elements of a bridge.

Hence time and money may be saved using the hydraulic lift because of the ease of operation, better position and working conditions for the inspector and faster inspections.

Disadvantages of the hydraulic lift include initial cost, possible blocking of traffic, number of personnel required to operate, and difficulty in reaching some areas of truss-type bridges.

A boat or barge can be used as a platform from which to measure scour with a leaded line, pole, or electronic device. They can be used as a means of reaching various bridge components, for example dolphins and fenders. They may act as a platform from which to climb so as to be able to reach and inspect the tops of various components again, such as dolphins, and as a safety measure when individuals are inspecting bridge elements over water, since the boat can assist significantly in rescue operation should an inspector have the misfortune of falling into the water. Larger boats may have a scaffolding or a frame construction on the boat to facilitate easier and safer inspection of those bridge elements under the deck portion.

Diving equipment of many different types, from scuba equipment to the type of equipment that requires a source of surface air, may be required occasionally. The end result is the same in that the diver has an outside source of air and is thus able to remain safely under water for rather extensive periods of time.

Scaffolding is a temporary, elevated structure used to support the inspector. Care should be used when constructing the scaffolding to make sure it is anchored securely and that it is strong enough to support the intended load. Hanging scaffolding should be checked for safe capacity with 4 times the intended load. Scaffolding requires considerable time to construct and take down; consequently, a snooper, if available and practical, is perhaps a much better method. Some types of scaffolds may be floated under a bridge and raised with a block and tackle.

The snooper is a special truck-mounted platform attached to hydraulic booms that provide special capability to the inspector. The boom system is designed to go under the superstructure for inspection while the truck is on deck. Figure 10.1 shows a typical snooper vehicle and the various booms and capability of moving the inspector to various locations under the superstructure. Figure 10.1 also shows the range of the booms and the capacity chart for the snooper vehicle used by the New Mexico State Highway Department.

The snooper should be thoroughly checked before each job. The inspectors should read carefully the instruction manual for the vehicle and use the safety features provided at all time.

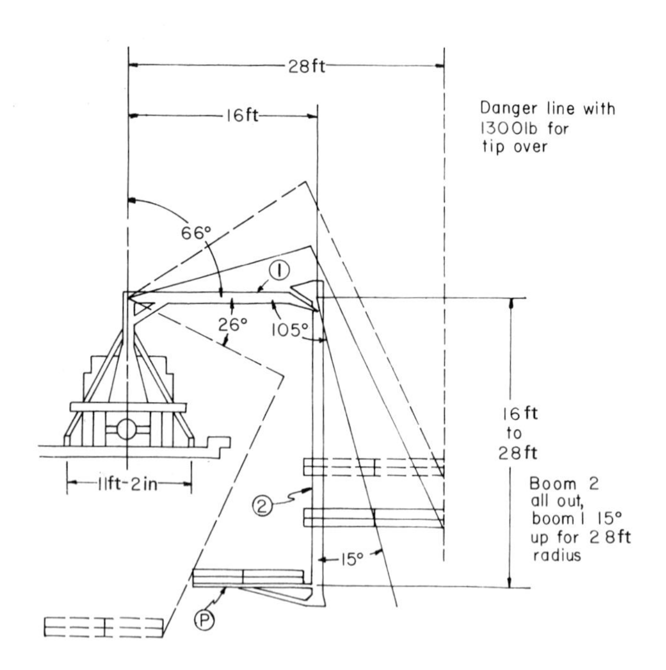

CAPACITY CHART (85% tipping factor)

RADIUS (ft.)	5	10	16	20	24	28
LOAD (lbs)	8500	4100	2400	1900	1600	1300

$$\text{TIPPING FACTOR} = \frac{\text{LOAD ON MACHINE}}{\text{TIPPING LOAD}}$$

YIELDING LOAD: 600lb at end of platform with factor of safety of 3

Fig. 10.1 Typical Boom Locations of a "Snooper Truck" and Corresponding Capacity Chart

C. Nondestructive Testing

The equipment discussed in this section is for some of the more sophisticated methods of nondestructive testing and inspection. These methods may be employed to gain in-depth information relative to a potentially critical condition discovered by the inspector, but which is beyond his capability to evaluate with the limited tools available to him. Some of these "sophisticated" methods of test and inspection include: (1) ultrasonics, (2) radiographics, (3) magnetic particle, (4) strain measurements. Each of these methods will be discussed briefly. However, it should be emphasized that the information presented here is only for the purpose of familiarizing the inspector with these methods and the capabilities of each, so that he will be informed about when to make recommendations concerning techniques.

1. Ultrasonic Testing

The term ultrasonic means vibrations or sound waves whose frequencies are greater than those which affect the human ear; i.e., frequencies which are greater than about 20,000 cycles per second. Typical ultrasonic equipment operates at frequencies anywhere between 500,000 cycles per sec. and 5,000,000 cycles per sec. In ultrasonic testing, a beam of energy travels through material such as steel with little loss. Upon striking a discontinuity, a portion of the energy is reflected much in the way that light is bounced back from a smooth surface. Radar operates on much the same principle.

Ultrasonic test equipment is capable of locating both surface and subsurface defects in metals including cracks, slag or other inclusions, segregation and lamination, pores or gas pockets, flaking, incomplete weld penetration and weld fusion, differences in structural properties of the metal and measuring the thickness of the metal from one side.

Ultrasonic test equipment is well suited for analyzing possible defects in steel bridge elements and connection points. The equipment itself is relatively simple and portable; however, a skilled operator is required to ensure the equipment is properly calibrated, and that signals are accurately interpreted.

2. Radiography

Radiographic inspection employs the use of x-rays (an electron beam) or gamma-rays (an energy beam from radioactive material such as radium) to detect defects in welds found on bridge structures. The energy beams of x-rays may vary from 50,000 volts to 2 million volts depending on the thickness of the metal being examined. A 2 million volt beam will penetrate metal up to 8 in. thick.

Where gamma-rays are used, the energy of the beam is constant and depends upon the gamma source being used. Besides radium, new radioactive materials such as cobalt, irridium, and cessium are used. These sources supply energy at given levels ranging from 250,000 to 2 million volts equivalent. Both systems employ a sensitive film placed on the opposite side of the joint or connection being inspected. The film records variations in the material by darkening of the film and this can be interpreted to determine if faults exist in the joint or connection.

Both x-ray and gamma-ray inspections are capable of locating surface as well as subsurface defects. Their use is primarily aimed at the detection of subsurface defects such as:

Cracks

Incomplete weld fusion

Slag and other inclusions

Incomplete weld penetration

Pores or wormholes (gas pockets)

In addition, both provide a permanent record of the condition of the area inspected in the form of the developed film.

X-ray equipment is available in a portable configuration which is capabile of x-raying steel up to 2 in. It requires a power source and a cooling system in addition to the x-ray tube and film.

The principle advantage of gamma-rays is that the energy source is available in a combination storage and projection device which is hand transportable. This device permits the radioactive material enclosed to be easily changed from a stored condition to an emitting condition by operation of a simple control.

Both the x-ray system and gamma-ray system have a common drawback. Both represent health hazards from radiation. In addition, both systems require well-qualified technicians for proper operation and personnel skilled in the interpretation of the information recorded on the films.

3. Magnetic Particles

Magnetic particle inspection is a method of inspection which sets up a magnetic field in the area of the metal to be inspected by passing electrical current from one prod or electrode to another. As the current passes through the metal from one prod to the other, a magnetic field is set up which is perpendicular, or at right angles to, the line between the prods. When a fine magnetic powder is applied to the surface to be inspected, the powder forms along these magnetic lines. If there is no defect, the pattern is uniform. If there is a defect, such as a crack along a weld between the prods, the defect will be outlined by the magnetic powder.

Magnetic particle inspection is, for the most part, limited to the detection of surface defects. However, an experienced interpreter of magnetic particle patterns may detect subsurface defects to a depth of about 1/2 in. A surface crack along the line between the prods may be detected readily, but a crack which is at right angles to the prod's axis may go undetected. Therefore, it is desirable to orient the prods in more than a single direction when checking for defects.

Portable equipment is available for the inspection of bridge structures. Such equipment is capable of operating on either 110 volts AC or a 12-volt automobile battery. All that is required for magnetic particle testing is the power source, the prods and the magnetic particles. It is essentially a two-person operation in which one holds a prod, applies the magnetic power and interprets the magnetic pattern produced while the second, or helper, moves the other prod.

4. Strain Measurements

The bonded resistance-strain gage developed in 1938 is said to be the most important single tool available to the engineer involved in the analysis of bridge stresses. The principle of the electrical resistance-strain gage is simple. Within the elastic limit for any material, the amount of strain (deformation) is proportional to the amount of stress. Thus, if one can measure the amount of deformation and the relationship between the strain and stress for the particular material is known, then one can determine the amount of stress. The bonded resistance-strain gage, about the size of a postage stamp, measures the amount of deformation in the metal to which it is bonded. This measure of deformation is transmitted electrically to recording instruments where stress computations are made.

The strain gage is capable of providing accurate information about the magnitude, distribution, and direction of the strains in loaded bridge structures. From these measurements, the internal stresses can be determined.

The equipment required for a stress analysis by means of strain gages can be relatively simple or complex, depending upon the degree of analysis to be performed. Basically, for each single point to be analyzed, a strain gage, electrical lead, and a recording mechanism is required.

Chapter 11

BRIDGE CAPACITY RATING

The engineering profession has given considerable attention to the design of bridge structures. Such designs have required new analytical techniques and specifications to ensure safety of the structure. Such designs offer the highway agencies new challenges and responsibility. However, these agencies have a greater responsibility in maintaining the safety of these new structures and, more important, those bridges which serve the public and have been in service for many years.

Many early bridges still exist throughout the states and are quite serv-icable when properly maintained. If the various transportation agencies were to replace these bridges, the cost would be prohibitive. Thus, proper maintenance and rating of these bridges is a necessity.

The rating of such bridges is performed by a combination of field inspection of the bridges and an analytical study, as guided by the AASHTO Manual for Maintenance Inspection of Bridges [1] and the Interim Specifications for Bridges 1976 [2]. These manuals provide guidelines for inspection of existing bridges, records of such bridges, ratings, and specifications for checking capacities.

The application of these guidelines will be illustrated through presentation of the rating of several types of bridges common in many states.

The AASHTO specifications as found in the Manual for Maintenance Inspection of Bridges 1974 [1], the Interim Specifications for Bridges 1976 [2], and the Standard Specifications for Highway Bridges [5], will be used to determine the allowable stresses in bridge elements. The age of the bridge must often be considered in determining an allowable stress. The allowable stress is usually taken as the stress used in the design of the structure or the stresses recommended by AASHTO at the time of construction.

There are two levels of capacity recommended by the AASHTO specifications at the present time. The inventory rating is defined as the load which produces a stress in the critical bridge element of 0.55 times the yield stress or the allowable stress used in design. The operating rating is defined as the maximum load that should be allowed on a bridge under any circumstances and should not exceed 0.75 times the yield stress or 75/55 times the allowable stress used in design.

The above definitions are fairly easy to apply to timber or steel structures. However, the determination of the capacity of a concrete structure or structural element is more involved. The load factor method is probably the simplest method in determining the safe load-carrying capacity. This method is described in the AASHTO Interim Specifications for Bridges 1976 [2].

In particular, the rating factors are:

1. Inventory level:

$$R.F._{(inv)} = \frac{M_u - 1.3M_D}{1.3(\frac{5}{3}M_{L+I})}$$

2. Operating level:

$$R.F._{(opr)} = \frac{M_u - 1.3M_D}{1.3(M_{L+I})}$$

where

R. F. = rating factor

M_u = ultimate moment capacity of beam

M_D = moment created by dead load

M_{L+I} = moment created by rating vehicle load + impact (e.g., HS20 truck)

$$M_D = \frac{wL^2}{8}, \text{ for simple spans}$$

where

w = unit dead load

L = length c-c bearing

The rating is determined by multiplying the rating factor by the standard load number. For example, if the standard load were an HS20 truck, the rating would be the rating factor times 20.

The typical rating vehicle load is the standard HS20 truck as defined in the Standard Specifications for Highway Bridges [5]. One positions the truck for maximum moment, computes the moment, increases it by the impact factor for the bridge, and obtains M_{L+I}.

M_u is defined as the maximum strength of the section. Methods for computing M_u can be found in the 1974 interim revision of the Standard Specifications for Highway Bridges [5]. A simplified form is:

$$M_u = 0.9A_s f_y (d - \frac{a}{2})$$

where

A_s = total area of tension steel

f_y = yield stress of steel

d = distance from tension steel to compression side of beam

a = depth of equivalent rectangular stress block

This expression is determined as follows:

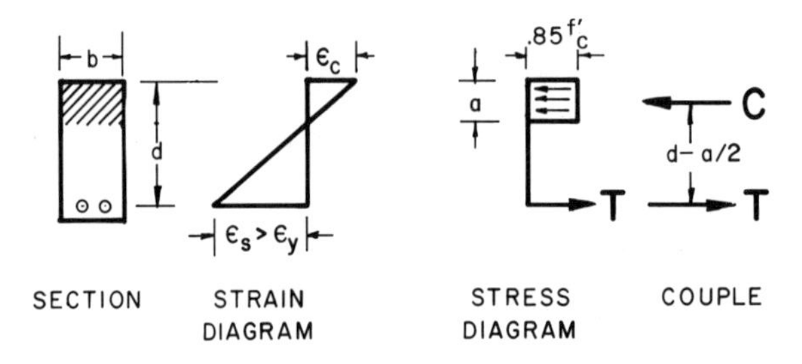

where

C = total compressive force

T = total tensile force

At failure:

$\epsilon_c = 0.003$ (ϵ = strain)

$\epsilon_s > \epsilon_y \quad \therefore \quad f_s = f_y$ (f = stress)

and

$T = f_y A_s$

but

$C = T$ and $C = 0.85f'_c ba$

so

$$a = \frac{T}{0.85f'_c b}$$

noting the couple diagram,

$$M = T(d - \frac{a}{2}) \quad \text{or} \quad A_s f_y (d - \frac{a}{2})$$

and the AASHTO equation is:

$$M_u = 0.9A_s f_y(d - \frac{a}{2})$$

which includes a reduction factor of 0.9.

This equation is good for normal, single reinforced beams or double reinforced beams if the compression reinforcement is neglected, which is common practice in analysis of this type. The strains shown could yield a definite max ρ for a particular combination of steel and concrete [3].

The general procedure for determining the load capacity of a bridge is as follows:

1. The total load-carrying capacity for the bridge is determined using the cross-sectional properties and the proper stress-rating level.
2. The dead load is determined for the structure.
3. The available live load capacity is determined by taking the difference between the total load capacity and the dead load.
4. The required live load capacity is determined for the structure using an AASHTO standard design vehicle, usually the HS20 truck load.
5. The ratio of the available live load capacity and the required live load capacity of the standard design vehicle times the standard vehicle load gives the equivalent capacity rating of the structure.

According to the AASHTO Manual for Maintenance Inspection of Bridges, 1970 [4], the weaker element of most bridges is the superstructure, not the piers and abutments. For this reason, the capacity rating of a bridge is determined from an analysis of the superstructure unless the inspection report or unusual structural configurations warrant analysis of the substructure. The capacity rating analysis usually includes the deck, stringers, floor beams, and trusses or girders depending upon the type of bridge. The following problems will include examples in each of these areas.

I. Deck Capacity Ratings

A. Example 1: Timber Deck

A timber deck was constructed of 2 × 4 timbers on edge over 8 × 15 timber stringers spaced at 3 ft. A 3 in. asphalt overlay is presently on the deck. An HS20 truck is to be used as the rating vehicle.

1. Total Moment Capacity of Deck

$$M_t = \text{(section modulus)} \times \text{(allowable stress level)}$$

S_x = section modulus = $\frac{1}{6}bd^2$ (rectangular section)

b = 1 ft width typically assumed

$d = 3\frac{5}{8}$ in. (nominal depth)

$S_x = \frac{1}{6}(12)(3.625)^2 = 26.28 \text{ in.}^3$

f = allowable bending stress = 1800 lb/in.^{2} (inventory)

= 2400 lb/in.^{2} (operating)

$M_t = 26.28(1800) = 47306$ in. • lb

$= \frac{1}{12}(47,306) = 3942$ ft • lb (inventory)

$= 3942\left(\frac{2400}{1800}\right) = 5256$ ft • lb (operating)

2. Dead Load Moment

$$M_D = \frac{1}{8}w_D(S_F)^2$$

S_F = clear spacing between stringers

w_D = total dead load on deck

Timber weight $= \frac{3.625}{12}(50 \text{ lb/ft}^3) = 15 \text{ lb/ft}^2$

Asphalt weight $= \frac{3}{12}(144 \text{ lb/ft}^3) = \underline{36 \text{ lb/ft}^2}$

$w_D = 51 \text{ lb/ft}^2$

$S_F = 36 - 8 = 28$ in.

$$M_D = \frac{1}{8}(51)\left(\frac{28}{12}\right)^2 = 35 \text{ ft} \cdot \text{lb}$$

3. Available Live Load Moment Capacity

$M_L = 3942 - 35 = 3907$ ft • lb (inventory)

$= 5256 - 35 = 5221$ ft • lb (operating)

4. Required Live Load Moment Capacity

$$M_r = \frac{PL}{4}$$

where

$P = \frac{1}{2}$ of the axle load (one wheel load)

$$L = S_F - \frac{T}{12}$$

T = 20 in. for HS20 vehicle

$$M_r = \frac{1}{8}\ (\text{axle load})\ (S_F - \frac{T}{12})$$

$$= \frac{1}{8}\ (32,000)\ (2.333 - \frac{20}{12})$$

$$= 2670\ \text{ft} \cdot \text{lb}$$

5. Safe Load Capacity Rating

The safe load capacity rating is determined from the equation:

$$R = \frac{M_L}{M_r} \times 20$$

where 20 represents the HS20 standard rating vehicle.

$$R_{inv} = \frac{3907}{2670} \times 20 = 29$$

The inventory rating is an HS29 vehicle for the deck.

$$R_{opr} = \frac{5221}{2670} \times 20 = 39$$

The operating rating is an HS39 vehicle for the deck.

B. Example 2: Concrete Slab

A concrete deck or concrete slab bridge that is simply supported requires a similar analysis. For this reason, an example of a slab bridge will be presented to illustrate the rating method for either a deck or slab bridge.

Find the inventory and operating capacity ratings for the structure shown.

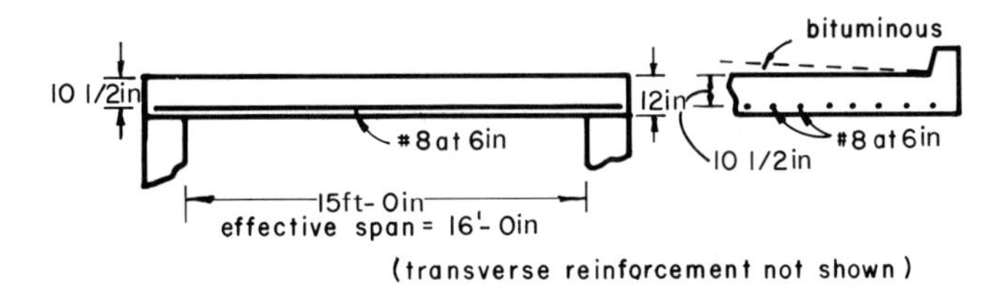

Clear span 15 ft

Rating vehicle: HS20 truck

Concrete strength, $f'_c = 3000$ lb/in.^{2}

Grade 40 steel ($f_y = 40,000$ lb/in.^{2})

Dead load:

$$\text{Concrete } \frac{12}{12}(150) = 150 \text{ lb/ft}$$

$$\text{3 in.-asphalt } \frac{3}{12}(144) = \underline{36} \text{ lb/ft}$$

$$w = 186 \text{ lb/ft (assuming 1 ft strip)}$$

$$M_D = \frac{WL^2}{8} = \frac{186(16)^2}{8} = 5952 \text{ ft-lb}$$

from AASHTO truck tables; (HS20 truck) [4]

$$M_{L+I} = 1677 \text{ ft-lb/ft}$$

A #8 bar has an area of 0.79 in.^{2}, and so

$$A_s = 2(0.79) = 1.58 \text{ in.}^2\text{/ft}$$

$$T = A_s f_y$$

$$= (1.58)(40,000)$$

$$= 63,200 \text{ lb}$$

$$a = \frac{T}{0.85 f'_c b}$$

$$= \frac{63.2}{0.85(3)(12)} \quad \text{(Use kips instead of pounds, b = 12 in. or 1 ft)}$$

$$= 2.06 \text{ in.}$$

$$M_u = 0.9(A_s f_y)(d - \frac{a}{2})$$

$$= 0.9(63.2)\ (10.5 - \frac{2.06}{2})$$

$$= 0.9(63.2)(10.5 - 1.03)$$

$$= 538.5 \text{ kips} \cdot \text{in.}$$

$$= 44.9 \text{ kips} \cdot \text{ft (maximum moment capacity)}$$

$$R.F._{(inv)} = \frac{M_u - 1.3M_D}{1.3(5/3M_{L+I})} = \frac{44.9 - 1.3(5.95)}{1.3(5/3)(16.77)}$$

$$R.F._{(inv)} = 1.02$$

$$R.F._{(opr)} = \frac{M_u - 1.3M_D}{1.3(M_{L+I})} = \frac{44.9 - 1.3(5.95)}{1.3(16.77)}$$

$$= 1.70 \ [\text{Note: } R.F._{opr} = \frac{5}{3}(R.F._{inv})]$$

$$R_{inv} = R.F._{(inv)}(HS20)$$

$$= (1.02)(HS20) = HS20.4$$

$$R_{opr} = (1.70)(HS20) = HS34.0$$

Bridge rating: inv: HS20.4; opr: HS34.0

II. Girder Capacity Rating

A. Example 3: Timber Beams

Consider a two-lane timber bridge that spans 20 ft (21 ft c-c of supports) with a 2 × 4 timber deck and 2 in. asphalt overlay. There are twelve 8 × 15 in. treated timber beams (stringers) at 2 ft spacing. The rating vehicle is an HS20 truck. The allowable stresses for the treated timber are 1600 lb/in.^{2} (inventory) and 2100 lb/in.^{2} (operating).

1. Total Moment Capacity of the Beams

$$M_t = f \times \frac{bh^2}{6} \times \frac{1}{12}$$

$$= (1600) \times \frac{8(15)^2}{6} \times \frac{1}{12} = 40{,}000 \text{ ft} \cdot \text{lb (inventory)}$$

$$= (2100) \times \frac{8(15)^2}{6} \times \frac{1}{12} = 52{,}500 \text{ ft} \cdot \text{lb (operating)}$$

2. Dead Load Moment

$$M_D = \frac{wL^2}{8}$$

$$\text{Weight of asphalt} = t \times S \times \gamma_a = \left(\frac{2\text{ in.}}{12}\right)(2)(144) = 48\text{ lb/ft}$$

$$\text{Weight of deck} = t \times S \times \gamma_t = \left(\frac{4\text{ in.}}{12}\right)(2)(50) = 33\text{ lb/ft}$$

$$\text{Weight of beams} = b \times d \times \gamma_t = \left(\frac{8\text{ in.}}{12}\right)\left(\frac{15\text{ in.}}{12}\right)(50) = \underline{42\text{ lb/ft}}$$

$$\text{Total weight } w = 123\text{ lb/ft}$$

where

t = thickness in in.

S = spacing in ft

γ = density in lb/ft^3

$$M_D = \frac{123(21)^2}{8} = 6780\text{ ft}\cdot\text{lb}$$

3. Available Live Load Moment

$$M_L = M_t - M_D$$

$$= 40,000 - 6780 = 33,220\text{ ft}\cdot\text{lb (inventory)/stringer}$$

$$= 52,500 - 6780 = 45,720\text{ ft}\cdot\text{lb (operating)/stringer}$$

4. Required Live Load Moment

$$M_r = \frac{1}{2}(M_{HS}) \times \frac{S}{4}$$

The value of M_{HS} is determined from the Table of Maximum Moments, Shears and Reactions—Loading HS20-44, Appendix A of the AASHTO Standard Specifications for Highway Bridges, 1977 [5].

$$M_{HS} = 168,000\text{ ft}\cdot\text{lb}$$

$$M_r = \frac{1}{2}(168,000)\left(\frac{2}{4}\right) = 42,000\text{ ft}\cdot\text{lb}$$

5. Safe Load Capacity Rating

$$R_{inv} = \frac{33,220}{42,000} \times 20 = 15.8$$

Inventory rating is an HS15.8 vehicle.

$$R_{opr} = \frac{45,720}{42,000} \times 20 = 21.8$$

Operating rating is an HS21.8 vehicle.

The inspector should note that since the required live load moment, M_r, was determined from an HS20, the rating factor, M_L/M_r, is multiplied by 20 for the HS rating. A similar result could be obtained using an HS15 for M_r and multiplying the rating factor by 15.

B. Example 4: Steel Beams

Consider a simple supported steel beam bridge with a concrete deck. The span is 25 ft. There are six W24 × 68 beams spaced at 7 ft 10 in. The yield stress for the steel is 36 kips/in.^{2} and the deck-beam action should be considered as noncomposite. Typically, the deck, overlay and beams will weigh between 1 and 2 kips per linear foot. Therefore, the assumed weight (dead load) for the structure is 1.5 kips/ft per beam.

1. Total Moment Capacity of the Beams

$$M_t = \frac{1}{12}(f_b) \times S_x$$

where S_x is the section modulus and is 153 in.^{3} for a W24 × 68 beam. The allowable inventory bending stress (f_b) for steel is $0.55F_y$ according to the AASHTO specifications [5].

$$f_b = 0.55(36) = 19.8 \text{ kips/in.}^2$$

$$M_t = \frac{1}{12}(19.8)(153) = 252.5 \text{ ft} \cdot \text{kips (inventory)}$$

The allowable operating bending stress if $0.75F_y$.

$$f_b = 0.75(36) = 27 \text{ kips/in.}^2$$

$$M_t = \frac{1}{12}(27)(153) = 344.3 \text{ ft} \cdot \text{kips (operating)}$$

2. Dead Load Moment

$$M_D = \frac{w_D L^2}{8}$$

$$= \frac{(1.5)(25)^2}{8} = 117.2 \text{ ft} \cdot \text{kips}$$

3. Available Live Load Moment

$$M_L = M_t - M_D$$

$$M_L = 252.5 - 117.2 = 135.3 \text{ ft} \cdot \text{kips (inventory)}$$

$$M_L = 344.3 - 117.2 = 227.1 \text{ ft} \cdot \text{kips (operating)}$$

4. Required Live Load Moment

$$M_r = \frac{1}{2}(M_{HS}) \times \frac{S}{5.5}$$

The moment produced by the rating vehicles (HS20 truck) is determined from the Tables for Maximum Moments, Shears, and Reactions—Loading HS20-44, Appendix of the AASHTO Specifications [4]. The distribution factor (S/5.5) is also determined from the distribution tables suggested for a concrete deck on steel stringers in the section on loadings in the AASHTO specifications.

$$M_{HS} = 282 \text{ ft} \cdot \text{kips}$$

$$M_r = \frac{1}{2}(282)\left(\frac{7.833}{5.5}\right) = 201 \text{ ft} \cdot \text{kips}$$

The rating of steel bridges must also include the effect of impact on the structures. The AASHTO specifications use the following expression in computing the impact as a percent increase.

$$I = \frac{50}{L + 125} \leq 30\%$$

The upper limit of 30% is based on experience from actual bridge results and research. For this particular structure,

$$I = \frac{50}{25 + 125} = 33\% > 30\%$$

Therefore, the impact is taken as 30%. The required live load plus impact is computed.

$$M_{r+I} = M_r \times (1 + I)$$

$$= 201 \times 1.30$$

$$= 261.3 \text{ ft} \cdot \text{kips}$$

5. Safe Load Capacity Rating

The inventory rating is computed as follows;

$$R_{inv} = \frac{135.3}{261.3} \times 20 = 10.4$$

The inventory rating is an HS10.4 vehicle. The capacity of this structure is considerably less than the present design standard of an HS20 truck. The bridge is probably old since the structure was noncomposite and the original design loads were probably much less than the present standards.

The operating rating is computed as follows;

$$R_{opr} = \frac{227.1}{261.3} \times 20 = 17.4$$

The operating rating is an HS17.4 vehicle. The operating capacity of this structure is less than the present design standards and therefore would normally be posted for reduced load capacity.

C. Example 5: Concrete Girders

The structure is a concrete deck girder bridge with 50 ft simple spans. The girder spacing is 5 ft 6 in. with a 29 ft roadway. The concrete is 3000 lb/in.^{2} strength and the reinforcement is Grade 40 steel. The rating vehicle is an HS20 truck. The capacity of the deck is determined using the technique described in Example 3. Determine the inventory and operating ratings.

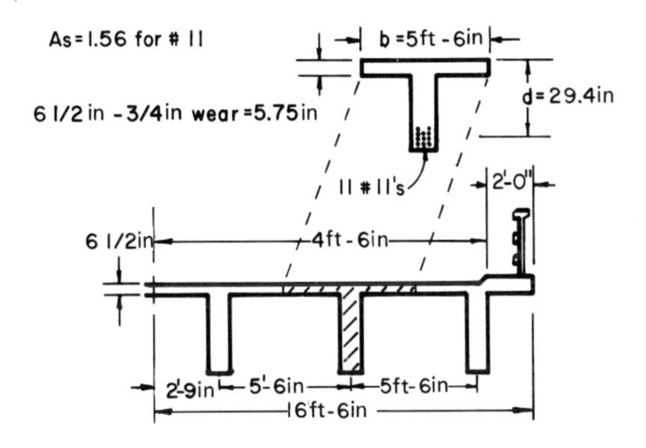

1. Total Moment Capacity of the Girders

If a is less than the effective thickness of the slab, one could treat the slab and stem as rectangular section for moment (this treatment is very common). Find a:

$$a = \frac{T}{0.85f'_c b}$$

$$T = A_s f_y$$

$$= 11(1.56)(40) = 686.4 \text{ kips}$$

$$a = \frac{686.4}{0.85(3)(66)} = 4.08 \text{ in.}$$

Since a is less than the slab thickness of 5-3/4 in., the treatment of the cross section as a rectangle is valid.

Find M_u:

$$M_u = 0.9(A_s f_y)\ (d - \frac{a}{2})$$

$$= 0.9(686.4)\ (29.4 - \frac{4.08}{2})$$

$$= 16,902 \text{ in.} \cdot \text{kips}$$

$$= 1408 \text{ ft} \cdot \text{kips}$$

2. Dead Load Moment

Weight of deck $= t \times b \times \gamma_c$

$$= (\frac{6.5}{12})\ (5.5)(0.150) = 0.45 \text{ kip/ft}$$

Girders $= b \times d \times \gamma_c$

$$= (\frac{12}{12})\ (\frac{36}{12})\ (0.150) \qquad = \underline{0.45 \text{ kip/ft}}$$

$$\text{Total} = 0.90 \text{ kip/ft}$$

Add 0.10 kip/ft for railings, sidewalks, or curbs; hence w_D = 1.0 kip/ft.

$$M_D = \frac{w_D L^2}{8} = \frac{1.0(50)^2}{8} = 312.5 \text{ ft} \cdot \text{kips}$$

3. Live Load Requires (from AASHTO <u>Bridge Specifications</u>)

$$M_{HS} = 628 \text{ ft} \cdot \text{kips}$$

$$M_L = \frac{1}{2} M_{HS} \times \frac{S}{6}$$

$$= \frac{1}{2}(638)\ \frac{5.5}{6} = 287 \text{ ft} \cdot \text{kips}$$

$$I = \frac{50}{L + 125} = \frac{50}{175} = 28.6\%$$

$$M_{L+I} = M_L \times (1 + I)$$

$$= 287(1.286) = 369 \text{ ft} \cdot \text{kips}$$

4. Determint the R. F. (Rating Factors) for Operating and Inventory Capacity

$$R.\ F._{opr} = \frac{M_u - 1.3M_D}{1.3(M_{L+I})} \quad \text{(operating)}$$

$$R.F._{inv} = \frac{3}{5}(R.F._{opr})$$

$$R.F._{opr} = \frac{1408 - 1.3(312.5)}{1.3(369)} = 2.09$$

$$R.F._{inv} = \frac{3}{5}(2.09) = 1.25$$

5. Compute the Capacity Ratings Based on HS20 Vehicle

$$R_{opr} = (R.F.) \times 20 = 41.8$$

The operating rating is an HS41.8 vehicle

$$R_{inv} = 1.25(20) = 25.0$$

The inventory rating for this bridge is an HS25 vehicle.

D. Example 6: Truss Bridge

In order to evaluate the proper capacity rating of a truss, all truss elements must be examined. This necessitates evaluation of the entire dead load of the truss and of influence lines for each truss element. The configuration of the various elements is shown in Fig. 11.1. The properties of these sections are as follows:

Properties

Members L_0U_1, U_4L_5, U_1U_4-End and top chords (Fig. 11.2)

i) Centroid from bottom of channel

Item	A	y	$A\bar{y}$
Top plate	1/4 × 12.25 = 3.06	6.125	18.75
Two channels	2 × 2.39 = 4.78	3.0	14.34
	7.84		33.09

$$\bar{y} = \frac{33.09}{7.84} = 4.2 \text{ in.}$$

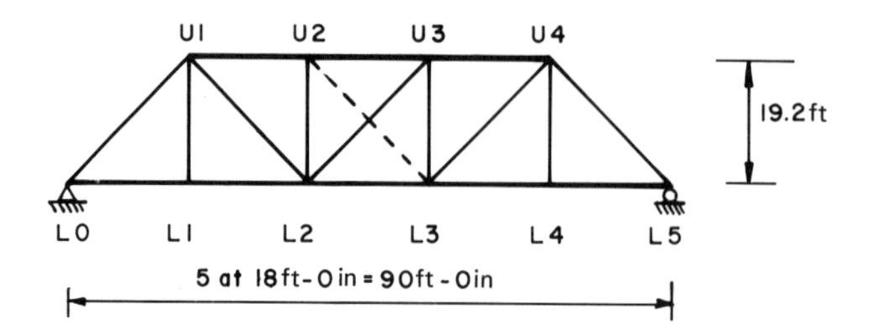

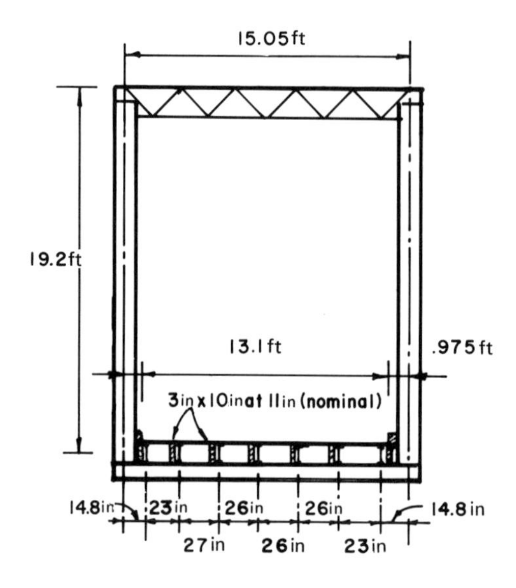

Fig. 11.1 Models for Truss Bridge—Example 6

ii) Inertia

Item	I_0	A	y	Ay^2
Plate	0.0160	3.06	1.91	11.11
Channels	26.0	4.78	1.22	7.12
	26.02			18.23

$I_T = 44.25 \text{ in.}^4$

$$r = \sqrt{\frac{I}{A}} = \sqrt{\frac{44.25}{7.8}}$$

$r = 2.38$ in.

$$(L/r)_{\text{end post}} = 26.3 \times \frac{12}{2.38} = 133$$

$$(L/r)_{\text{top chord}} = 18.0 \times \frac{12}{2.38} = 91$$

iii) Allowable stress from AASHTO Manual [4]

$$F_a = 10{,}000 - 0.152(L/r)^2$$

$$= 7.3 \text{ kips/in.}^2 \text{ (end post)}$$

$$= 8.7 \text{ kips/in.}^2 \text{ (top chord)}$$

iv) Weight (steel weight = 3.4 lb/ft in.^{2} area)

Top plate: $(\frac{1}{4} \times 12.25)(\frac{3.4 \text{ lb/ft}}{\text{in.}^2})$	10.4 lb/ft
2 channels	16.4 lb/ft
Batten plates	2.2 lb/ft
Say	29.0 lb/ft

Members L_0L_1, L_1L_2, L_3L_4, L_4L_5 — Bottom chords (Fig. 11.2a)

$$\text{Area} = 2 \times 2 \times \frac{5}{8} = 2.5 \text{ in.}^2$$

$$\text{Weight} = 2.5 \times 3.4 = 8.5 \text{ lb/ft}$$

Member L_2L_3 — Bottom chord (Fig. 11.1b)

$$\text{Area} = 2 \times 2.5 \times \frac{3}{4} = 3.75 \text{ in.}^2$$

$$\text{Weight} = 3.75 \times 3.4 = 12.76 \text{ lb/ft}$$

Member L_2U_1, L_3U_4 — End diagonal (Fig. 11.2c)

$$\text{Area} = 2 \times 2 \times \frac{1}{2} = 2 \text{ in.}^2$$

$$\text{Weight} = 2 \times 3.4 = 6.8 \text{ lb/ft}$$

Members L_2U_3, L_3U_2 — Center diagonal (Fig. 11.2d)

$\frac{7}{8}$ in. bar with turnbuckle

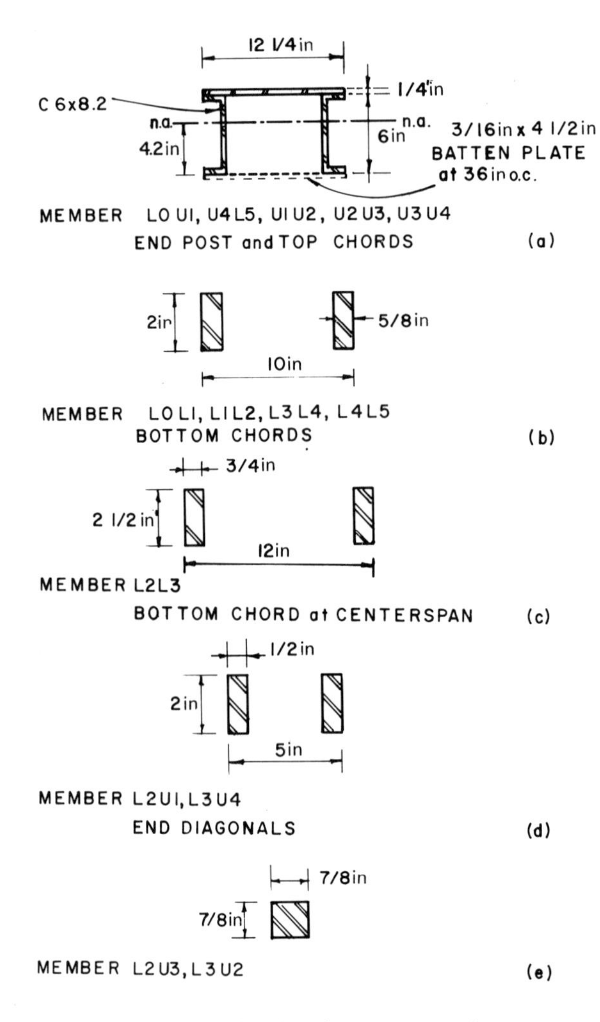

Fig. 11.2 Cross-Sectional Properties for Truss Bridge—Example 6

Area = 0.766 in.^{2}

Weight = 0.766 × 3.4 = 2.6 lb/ft

Members L_1U_1, L_2U_2, L_3U_3, L_4U_4— Verticals (Fig. 11.2e)

i) Inertia

A = 3.9 in.^{2}

$$I_x = 2 \times 7.4 = 14.8 \text{ in.}^4$$

$$I_y = 2 \times 0.48 + 3.9(3.24)^2 = 41.9 \text{ in.}^4$$

$$r = \sqrt{\frac{I}{A}} = \sqrt{\frac{14.8}{3.9}}$$

$$= 1.95 \text{ in.}$$

$$\frac{L}{r} = \frac{19.2 \times 12}{1.95} = 118.1$$

ii) Allowable stresses (compression)

$$F_a = 7.88 \text{ kips/in.}^2$$

iii) Weight

Two channels	2×6.7	= 13.4 lb/ft
Lacing bars		= 1.6 lb/ft
		15 lb/ft

Weight of the Truss

Members	No.	Length	Weight per foot	Total Weight
End post L_0U_1 and U_4L_5	2	26.318	30.00	1579.1
Top chord U_1U_4	3	18.000	30.00	1620.0
Bottom chord L_0L_1, L_1L_2, L_3L_4, and L_4L_5	4	18.000	8.50	612.0
Bottom chord L_2L_3	1	18.00	12.76	229.7
Members L_2U_1 and L_3U_4	2	26.318	6.8	358.0
Members L_2U_3 and L_3U_2	2	26.318	2.603	137.0
Vertical members L_1U_1, L_2U_2, L_3U_3, and L_4U_4	4	19.200	15.00	1152.0

Weight of truss = 5687.8 lb

Add 30% for connection = 1706.3 lb

Wooden curb $\frac{6 \times 6 \text{ in.}}{144} \times 50 \times 90 = 1125.0$ lb

Railing 10 lb/L. F. at 90 $= \underline{900.0}$ lb

Total dead load = 9419.1 (say 9500 lb)

Weight of the truss on each panel point

$9500 \div 5 = 1900$ lb

Weight of the deck, stringers, and
floor beam which is transferred by
floor beam at every panel point = 4515 lb

Total dead load at panel points

L_1, L_2, L_3 and L_4 = 1900 + 4515 = 6415 (Use P = 6.42 kips)

Dead load at panel point L_0 and L_5 due to the truss is

$1900 \div 2 = 950$ lb

Dead load due to deck, stringers, and floor beam

$4515 \div 2 = 2258$ lb

Applying these concentrated loads to the panel points gives the dead load bar forces, as shown in Fig. 11.3.

Live Loads

i) Influence-line diagrams

The resulting influence-line diagrams for each truss bar are shown in Fig. 11.4. These diagrams will now be used to evaluate maximum live load forces.

ii) Distribution of live loads

Two types of live loads are considered: a) truck load (H15) and b) lane load.

a) Truck load

The application of the H15 truck is positioned laterally on the bridge such as to induce the maximum reaction of the truss, as shown in Fig. 11.5. The maximum LL reaction is computed as:

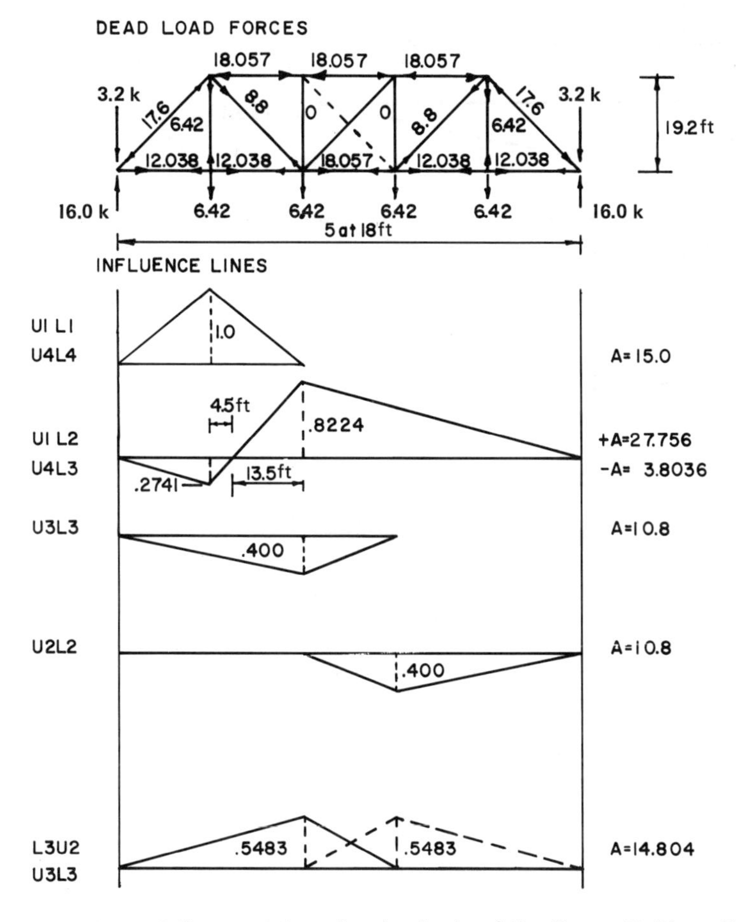

Fig. 11.3 Influence Lines for Analysis of the Truss Bridge—Example 6

$$\text{Reaction} = \frac{12(6.075 + 12.075)}{15.05} = 14.47 \text{ kips}$$

$$\text{Impact} = \frac{50}{90 + 125} = 0.2326$$

The total truss span of 90 ft is used here since the loads are used for truss analysis.

$$\begin{aligned} \text{L + I reaction} &= 14.47 \times 1.2326 \\ &= 17.84 \text{ kips (rear axle)} \\ &= 4.46 \text{ kips (front axle)} \end{aligned}$$

The resulting truck, which will be positioned along the truss, is shown in Fig. 11.6.

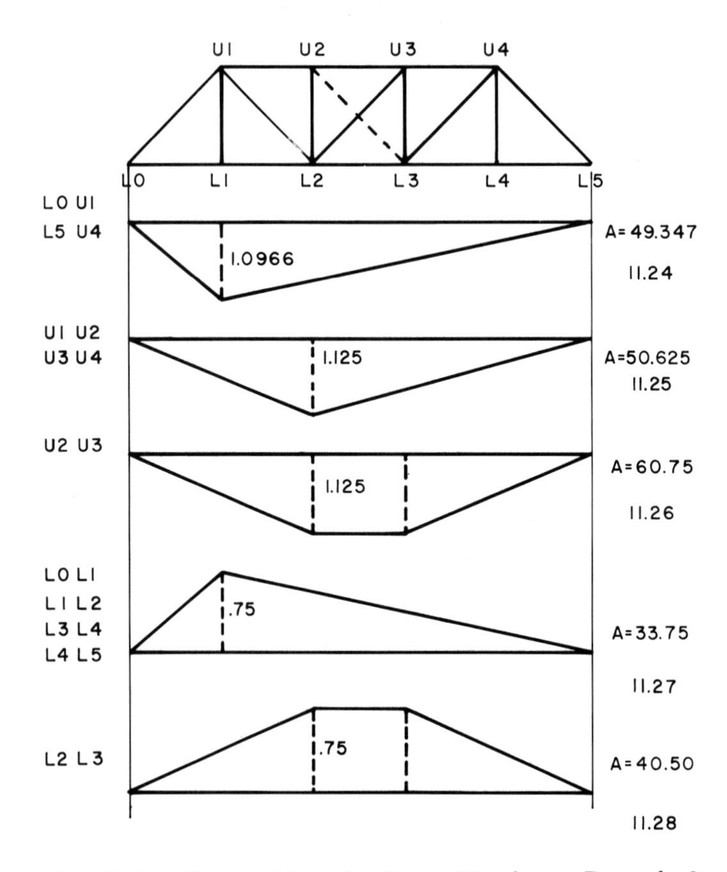

Fig. 11.4 Influence Lines for Truss Members—Example 6

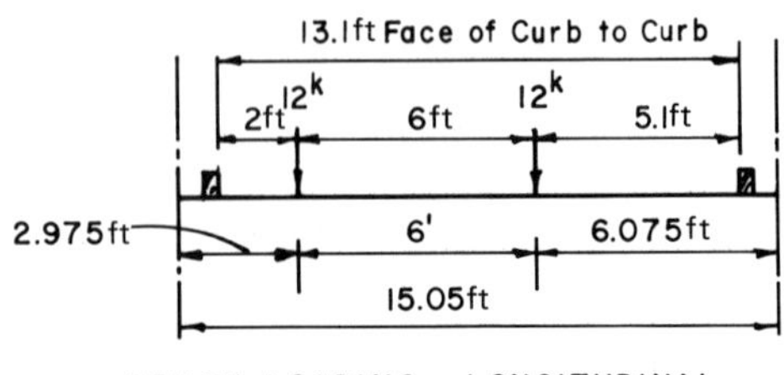

Fig. 11.5 Placement of Truck Load on Bridge Deck, Laterally

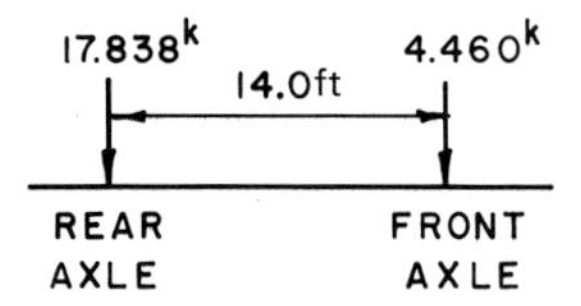

Fig. 11.6 Model of Truck Loading for Bridge, Longitudinally

b) Lane load

The uniformly distributed load over a 10.0 ft lane gives w = 480/10 = 48 lb/ft. The concentrated load for shear distributed over 10 ft width of lane is 19.5/10 = 1.95 kips/ft. The uniform load w is positioned as shown in Fig. 11.7. The maximum LL reaction is computed as:

$$\text{Reaction} = \frac{10w \times 9.075}{15.05} = 6.03w$$

$$\text{LL} + \text{I reaction} = 6.03w \times 1.2326 = 7.4325w$$

For the concentrated load

$$R = 7.4325 \times 1.95 = 14.493 \text{ kips acting on one truss}$$

For uniform load

$$F = 7.4325 \times \frac{8}{1000} = 0.357 \text{ kips/ft acting on one truss}$$

iii) Forces in various members

Applying the equivalent live loads (truck or lane) with the various influence lines gives the following resultant forces, as shown in Tables 11.1 and 11.2.

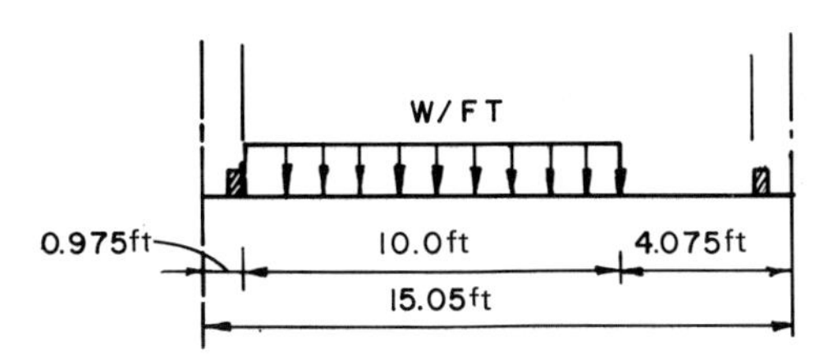

Fig. 11.7 Placement of Lane Loading on Bridge Deck—Example 6

Table 11.1 Forces in Truss Members Due to Truck Loading—Example 6

Member	Type	Force in the Member	Force
L_0U_1, L_5U_4	Comp.	17.838 × 1.0966 + 4.460 × 0.8838	-23.501
U_1U_2, U_3U_4	Comp.	17.838 × 1.1250 + 4.460 × 0.8333	-23.784
U_2U_3	Comp.	17.838 × 1.1250 + 4.460 × 1.1250	-25.085
L_0L_1, L_1L_2, L_3L_4, L_4L_5	Tension	17.838 × 0.7500 + 4.460 × 0.6042	16.073
L_2L_3	Tension	17.838 × 0.7500 + 4.460 × 0.7500	16.724
U_1L_1, U_4L_4	Tension	17.838 × 1.0000 + 4.460 × 0.2222	18.829
U_1L_2, U_4L_3	Tension	17.838 × 0.8224 + 4.460 × 0.6092	17.387
U_1L_2, U_4L_3	Comp.	17.838 × 0.2741 + 4.460 × 0.0609	-5.161
U_2L_2, U_3L_3	Comp.	17.838 × 0.4000 + 4.460 × 0.2444	-8.225
L_2U_3, L_3U_2	Tension	17.838 × 0.5483 + 4.460 × 0.3351	11.275

iv) Stress analysis

Summing the dead and live load forces for each member and the corresponding properties, the proper rating for each member is obtained as shown in Table 11.3.

Summary of Rating

The results of this study indicate the following:

i) Timber deck, not computed here but manner similar to Example 1.

a) Inventory rating—H4

b) Operating rating—H6

Table 11.2 Forces in Truss Members Due to Lane Loading—Example 6

Member	Force in the Member		Total Force
	Due to Uniform Load	Due to Concentrated Load	
L_0U_1, L_5U_4	0.357 x 49.347 = 17.617	14.493 x 1.0966 = 15.893	-33,510
U_1U_2, U_3U_4	0.357 x 50.625 = 18.073	14.493 x 1.125 = 16.305	-34,378
U_2U_3	0.357 x 60.750 = 21.688	14.493 x 1.125 = 16.305	-37.993
L_0L_1, L_1L_2, L_3L_4, L_4L_5	0.357 x 33.750 = 12.049	14.493 x 0.750 = 10.870	22.919
L_2L_3	0.357 x 40.50 = 14.459	14.459 x 0.750 = 10.870	25.329
U_1L_1, U_4L_4	0.357 x 15.00 = 5.355	14.493 x 1.00 = 14.493	19.848
U_1L_2, U_4L_3	0.357 x 27.756 = 9.909	14.493 x 0.8224 = 11.919	21.828
U_1L_2, U_4L_3	0.357 x 3.0836 + 1.101	14.493 x 0.2741 = 3.973	-5.075
U_2L_2, U_3L_3	0.357 x 10.80 = 3.856	14.493 x 0.400 = 5.797	-9.653
L_2U_3, L_3U_2	0.357 x 14.801 = 5.285	14.493 x 0.5483 = 7.947	13.232

Table 11.3 Stress or Loading and Capacity and Rating for Various Truss Members—Example 6[a]

Member	Dead load force, kips	Live load force, kips	Length, in.	Area
1. L_0U_1, L_5U_4	-17.600	-33.510	315.82	7.8425
2. U_1U_2, U_3U_4	-18.057	-34.378	216.00	7.8425
3. U_2U_3	-18.057	-37.993	216.00	7.8425
4. L_0L_1, L_1L_2, L_3L_4, L_4L_5	12.038	22.919	216.00	2.500
5. L_2L_3	18.057	25.329	216.00	3.75
6. U_1L_1, U_4L_4	6.42	19.848	230.40	3.90
7. U_1L_2, U_4L_3	8.800	21.828	315.82	2.00
8. U_1L_2, U_4L_3	8.800	-5.074	315.82	2.00
9. U_2L_2, U_3L_3	0.000	9.653	230.40	3.90
10. L_2U_3, L_3U_2	0.00	13.232	315.82	0.7656

Member	r, in.	L/r	Dead load stress, kips/in.^{2}	Live load stress, kips/in.^{2}	Allowable stress, kips/in.^{2}
1.	2.3752	133	-2.244	-4.273	7.311[b]
2.	2.3752	91	-2.302	-4.384	8.741[c]
3.	2.3752	91	-2.302	-4.845	8.741[c]
4.	—	—	4.815	9.168	12.500
5.	—	—	4.815	6.754	12.500
6.	—	—	1.646	5.089	12.500
7.	—	—	4.40	10.914	12.500
8.	—	—	4.40	-2.537	—
9.	1.95	118.2	0.000	2.475	7.878
10.	—	—	0.000	17.283	12.500

Table 11.3 Continued

Member	Available stress for live load, kips/in.^{2}	Inventory rating
1.	2.874	H10
2.	4.254	H14.6
3.	4.254	H14.6
4.	7.685	H12.6
5.	7.685	H17.1
6.	10.854	H25.6
7.	8.100	H11.1
8.	—	Dead load tension greater than live load compression
9.	7.878	H47.7
10.	12.500	H10.8

[a]Compression = -; tension = +.
[b]Note: Only 70% allowable stress is used for rating as S. E. top chord, L_0U_1 rusted and buckled.
[c]25% allowable stress reduction for rust.

ii) Steel stringers, not computed here, but in manner of Example 4.

a) Inventory rating—H1.4

b) Operating rating—H1.8

iii) Floor beams, not computed here, but in manner of Example 4.

a) Inventory rating—H8.0

b) Operating rating—H10.0

iv) Truss

a) Inventory rating—H10.0

b) Operating rating—H14.0

The bridge is, therefore limited by the stringers. If the deck system is improved, the rating could be substantially increased.

III. Bearing Capacity Rating

The most common problems in bridge bearings is the result of deterioration or damage of the bearings. This deterioration or damage usually results in a loss of contact area between the load-carrying member and the support. The safe load carrying capacity then becomes a function of the remaining contact area of the bearing device.

The allowable stresses in bearing are based on the AASHTO Bridge Specifications [4]. For timber, the allowable unit of stress is taken as compression stress perpendicular to the grain for bearing lengths of 6 in. or more. For bearing lengths between 3 and 6 in. the allowable stress is obtained by multiplying the compression stress perpendicular to the grain by the factor:

$$\frac{L + 3/8}{8}$$

where L is the length of bearing in inches along the grain of the wood.

The allowable bearing stress for steel is taken at 80% of the yield stress. The yield stress is normally determined by the age of the bridge, and typical ranges are available in the AASHTO Bridge Maintenance Inspection Manual 1974 [1].

The allowable concrete-bearing stress is normally taken as 30% of the 28-day compression strength of the concrete. When the supporting surface is wider on all sides than the loaded area, the allowable bearing of the supporting surface is increased by A_2/A_1, but not by more than 2. When the supporting surface is subject to high edge stresses due to deflection or eccentric loading, the allowable bearing stress is multiplied by 0.75.

To illustrate a method of determining the safe load capacity in bearings, a cast-in-place concrete T beam bridge is presented. This type structure is a bridge common across the Rio Grande in southern New Mexico. The bridge is multiple simple spans of 34 ft with the beam stems 12 in. wide and 20 in. deep at 7 ft spacing. The deck is 8 in. thick with diaphragms at the ends and at midspan. The original bearing length of the beams was 12 in., but many are now damaged to such an extent that the effective bearing depth is about 5 in. The supporting edges are subjected to deflection; hence the allowable bearing stress is

$$f_b = 0.30f'_c \times 0.75$$

The assumed concrete strenght is 3000 lb/in.^{2}, which yields a bearing resistance as follows:

$$R_b = f_b A_b$$

$$f_b = 0.30(3000)(0.75)$$

$$A_b = bL_b = 12 \times 5 = 60 \text{ in.}^2$$

$$R_b = (675 \text{ lb/in.})(60) = 40,500 \text{ lb}$$

The dead load on each beam bearing is computed as follows:

$$\text{Deck weight} = t \times S \times \gamma$$

$$= \frac{8}{12} \times 7 \text{ in.} \times 150 \text{ lb/ft}^3 = 700 \text{ lb/ft}$$

$$\text{Beam weight} = b \times h \times \gamma$$

where

b = stem width

h = stem height

w_D = total dead load

$$= \frac{12}{12} \times \frac{20}{12} \times 150 \text{ lb/ft}^3 = 250 \text{ lb/ft}$$

$$w_D = 700 + 250 = 950 \text{ lb/ft}$$

$$R_D = \frac{1}{2} w_D = \frac{1}{2}(950)(34) = 16,150 \text{ lb}$$

The available live load capacity is,

$$R_L = R_b - R_D = 40,500 - 16,150$$

$$= 24,350 \text{ lb}$$

The required live load for an HS20 truck is determined.using the AASHTO Bridge Specifications and the distribution equation:

$$R_{rq} = \frac{1}{2} \times \text{reaction} \times \frac{S}{6}$$

$$= \frac{1}{2} \times 52,200 \times \frac{7}{6}$$

$$= 30,450 \text{ lb}$$

The inventory rating is computed in the usual manner:

$$R_{inv} = \frac{24,350}{30,450} \times 20 = 16.0$$

The inventory rating is an HS16 vehicle.

The operating rating may be approximated by simply using the overload factor of 5/3.

$$R_{opr} = \frac{5}{3}(16) = 26.7$$

The operating rating is an HS26.7 vehicle. No posting would be required since the operating is greater than the design vehicle. These ratings do not reflect any redistribution through the end diaphragms.

References

1. American Association of State Highway and Transportation Officials, Manual for Maintenance Inspection of Bridges, 1974, Washington, D.C., June 1974.

2. American Association of State Highway and Transportation Officials, Interim Specification—Bridges, 1976, Washington, D.C., 1976.

3. C. K. Wang and C. G. Salmon, Reinforced Concrete Design, Intext Education, New York, 1973.

4. American Association of State Highway and Transportation Officials, Manual for Maintenance of Bridges, 1970, Washington, D.C., 1970.

5. American Association of State Highway and Transportation Officials, Standard Specifications for Highway Bridges, Washington, D.C., 1977.

Chapter 12

COMPUTER AIDED BRIDGE CAPACITY RATING AND EVALUATION

I. Introduction

Many of the structures that come under the Bridge Maintenance Inspection Program are complex and therefore difficult to analyze for safe load-carrying capacity. At the same time, the data generated in the inspection of bridges by any one agency quickly becomes voluminous. The computer can be a very useful aid in analyzing complex structures or in effectively using the great volume of information collected concerning the bridge system, if the proper computer software is available. Several software systems are available that aid in the computation of the safe load-carrying capacity of bridges, and at least one system is available for the evaluation of bridges along any route for overweight vehicles or permit loads.

II. Bridge Route Evaluation

The advent of larger and more accessible computer systems has led to development of software that gives the practicing engineer a means for getting immediate answers to routine questions. The computer program ØVLØAD is one such program. ØVLØAD has the capability of automatically checking potential overload situations against the capacity of every bridge along a proposed route. Such comparisons enable the engineer to make rational decisions regarding issuance of permits for overweight vehicles.

A. Background

A permit is usually necessary if, among other criteria such as length and width, the gross weight exceeds 80,000 lb (357 kN). Any truck requiring a permit because of gross weight is referred to bridge engineers for evaluation of the effects on the highway and bridge structures. Single-trip permits in New Mexico for overweight vehicles have been requested for vehicles with gross weights over 2.56 MN (800,000 lb) and multiple-trip permits (sometimes called blanket or annual permits) have been requested for vehicles with gross weights of 712 kN (160, 000 lb).

Both types of requests have their own implications with regard to potential damage. A single-trip permit implies an infrequent load and should be properly compared to an operating capacity as defined in the Manual for Maintenance Inspection of Bridges [1]. The operating rating stress is the absolute maximum permissible stress level to which a structure may be subjected. Special permits for vehicles heavier than legal load are issued only if such loads are distributed so as not to produce stress in excess of the operating rating stress level.

A multiple-trip permit load should be compared with the inventory rating load. The inventory rating load is designated as that load which can utilize an existing structure for its design life without appreciable deterioration of the bridge. Exceeding an inventory rating by a large number of times invites the possibility of fatigue failure or other types of accelerated deterioration.

A load exceeding the inventory rating but not the operating rating will not cause the bridge to collapse but will shorten the life of the structure by an indefinable amount. Consequently, it is of particular importance that the engineer be able to make a comparison of the load distribution and magnitude of the overweight vehicle with the capacity ratings of all bridges along a proposed route. Often these permit requests take engineers away from their normal duties when they are required to pinpoint potential problem bridges and reanalyze them for the particular overload vehicle. The procedure is not only time consuming, taking up valuable engineering man hours, but has an increased potential for serious error.

First, it is possible that a potentially dangerous structure could be overlooked. Second, a rush analysis increases the chance for error. Also, the data available does not always reflect the current condition of the bridge.

In 1972, when the problem seemed to be becoming worse, the New Mexico State Highway Department (NMSHD) decided that a more automated method of operation was desirable. Two facts stood out. First, a computerized technique for loacting bridges along a proposed route as well as performing a structural analysis would certainly speed up the operation and allow the engineer a decision on each new overweight vehicle considered. Also, the latest structural conditions of all bridges on the state system were readily available through the Bridge Maintenance Inspection Program. The inventory and operating ratings, as well as other pertinent data such as bridge type and location, were already stored on magnetic tape. The civil engineering department at New Mexico State University, developed a computer system that met the criteria of the NMSHD to streamline the permit operation by utilizing the bridge maintenance inspection information [2, 9].

The bridge inspection inventory data bank includes the results of a complete analysis of every bridge in New Mexico. Both the inventory and operating ratings for each bridge were recorded as an AASHTO [3] HS loading.

All that remained was a method to couple this data with a system to provide an equivalent HS loading for any potential configuration of axle loads and spacing.

A new analysis of each structure along each route was rejected as impractical. Detailed structural data was not available from the bridge inspection inventory system. Also, a software system that analyzed every bridge would require large quantities of computer time and space. Finally, since each bridge was already analyzed and given a safe load capacity rating as part of the continuing bridge inspection program, a new analysis would be repetitious. With these facts in mind, the following method of equivalent loading was developed.

B. Method of Equivalent Loading

Since bending moment is considered to be the most likely mode of failure in a typical bridge, it was decided to use bending moment as the common denominator for relating a proposed truck to an HS truck. Hence a typical matrix analysis technique was developed to determine the maximum moment produced by an HS20 truck and the proposed overweight vehicle when both were driven across a particular bridge. An equivalent HS rating is assigned to the overweight truck by the ratio of the two bending moments times 20. This equivalent HS rating can be compared directly with the HS ratings which have been previously computed for each bridge.

Careful study shows such a comparison can be developed for several types of bridges. For instance the moment ratio would suffice for the main truss of a bridge by using the span of the truss. There exists a direct relationship between the forces in the chord members of a truss and the bending moment produced by a load on an equal span. Stringers in the truss could be evaluated by creating an equivalent HS loading based on a ratio of moments created by the HS20 truck and the overweight vehicle on a simple span of a length equal to the panel size of the truss. Similarly, a ratio of reactions of the stringers due to the two trucks produce an equivalent HS truck based on the floorbeam strength.

The analysis produces moments, shears, and reactions, although most bridges are compared by moment ratio alone. Most load capacities in the data bank were computed by bending moment. It should be noted that even if shear is critical in a simple bridge, the ratio of shears would be of similar magnitude to the ratio of moments and therefore an equivalent HS truck of similar magnitude would be produced.

This equivalent HS loading could be compared with either the inventory rating or the operating rating depending upon the circumstances of the permit request. A program was written to automatically compute this equivalent loading for each particular bridge and compare it with the rating of the particular bridge.

C. Description of Program

The program ØVLØAD consists of the main program and three subroutines. The main program receives input data, reads stored data on bridges, determines whether or not a bridge is on the requested route, makes a comparison between safe load capacity and the required capacity via equivalent loading, and prints information on inadequate bridges. The subroutines compute the equivalent HS loading for a given overweight vehicle on each particular bridge. A simplified flow chart of the software program is shown in Fig. 12.1.

The process is initiated by the input of the route numbers and Department of Defense (DOD) numbers of the route over which the proposed overloaded truck is to travel or the route numbers and inclusive mile posts. The axle spacings and axle loads of the truck are input for use in computation of an equivalent HS truck. Finally, whether the comparison will be with inventory rating or operating rating is entered. This input is via a cathode-ray tube (CRT) terminal located in the office of the bridge maintenance engineer where the overweight vehicle permits are evaluated. Access to the computer and program is immediate.

All bridge data are stored by bridge number on magnetic disks. The data for each bridge are read and evaluated with regard to route and DOD section. Any bridges not on the desired route are rejected before any equivalent load computations are made. If the bridge is on the desired route and within the desired DOD sections, the structure is classified as one of four types and an equivalent loading determined.

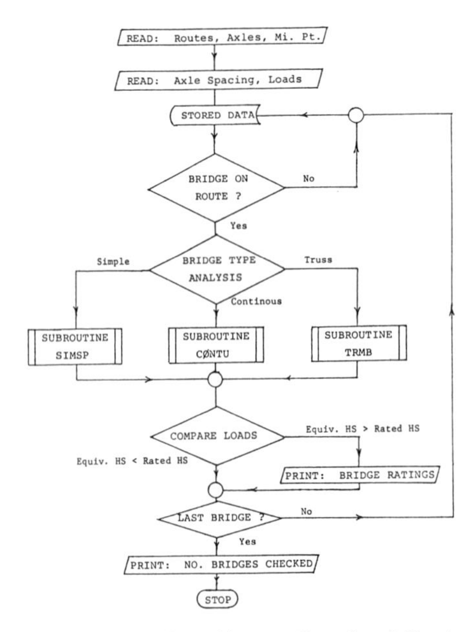

Fig. 12.1 Flow Chart—Overload Route Program

The bridge structures in New Mexico can be divided into four general categories: simple span bridges, continuous span bridges, truss type bridges, and concrete box culverts. Although other types of bridges exist, such as rigid frame structures, each can be classified into one of the above categories for the computation of an equivalent HS truck rating of an overweight vehicle.

All types except the concrete box culvert (CBC) use a similar computation procedure. The CBCs do not have an equivalent HS truck determined. Instead, since deadload overburden makes the equivalent load method impractical, the rated capacity is printed without comparison at the operator's request.

The general procedure utilizes the span lengths and/or panel lengths of the bridge to be checked and the axle spacing and axle loads of the overweight vehicle. The first axle of the vehicle is placed at a critical location of the first span and the critical bending moments are computed. These moments are retained for later comparisons. The next axle is moved to the critical location, and new bending moments are computed. The new bending moments are compared to the previous bending moments, and the larger values are retained. This process continues until all positions of the vehicle have been compared. The maximum bending moments produced by a standard HS20 truck are computed in a similar manner for the same span. An equivalent HS rating is assigned to the overloaded truck by the ratio of bending moments produced by the two trucks times 20. This equivalent HS value is stored for later comparison. The overload vehicle is then moved to the next span. The entire procedure is done again for this span. The new equivalent is compared to the old and the larger value retained. This procedure is continued until all spans have been checked. The largest equivalent HS rating is then returned to the main program for comparison with either the operating or inventory rating as requested.

For simple spans, positive moments and midspan are compared. For continuous bridges, comparisons are made of both positive moments and negative moments in each span. Some bridges combine simple spans and continuous spans. All are checked, comparing the equivalent HS loading to the rated HS capacity.

Truss bridges seemingly present a more difficult problem. However, comparisons of moment in stringer spans, reactions of stringers, and moments of the truss span provide logical and easily obtainable equivalent loads for comparison. Each different equivalent load is carried back to the main program for individual comparison with the safe load capacity.

The equivalent HS loads are compared with the safe load capacity, either operating or inventory rating, which had been previously assigned to that structure by a complete analysis. If the equivalent HS rating is greater than the stored capacity, the bridge is considered inadequate, and pertinent information about the bridge is shown on the CRT terminal and stored to be printed on a separate print-out. The information gives the user a quick identification of critical bridges, possible need for additional analysis, or whether detours might be necessary or available. This procedure is repeated until every bridge on the designated route has been checked and every inadequate bridge listed.

A typical output is shown in Fig. 12.2. Each bridge number, route, DOD section, and direction of the route is listed. This information is followed by the type of bridge, a description of the location of the bridge, the critical span length, the equivalent HS load capacity required and the rated HS load capacity. A ratio of equivalent load to the rated load is printed to assist in evaluating the bridge. Trusses have three equivalent HS truck printouts. The first indicates an equivalent truck based on stringer moments, the second indicates an equivalent truck based on floor beam reaction loadings, and the third indicates an equivalent truck based on overall moment.

In all cases the operator has an immediate evaluation via the CRT terminal and a permanent record via the high speed printer. The immediate readout is especially useful if there are no critical bridges. Any critical bridges must be reviewed by an engineer before any decision is made on the issuance of a permit.

Although the analytical procedures used are relatively simple and contain approximations such as assuming prismatic members and no lateral distribution of load, these simplifications and assumptions were essential to the successful mating of ØVLØAD to the bridge inspection program data. This successful mating was accomplished only through a thorough understanding of the field operation, data collection, safe load rating techniques, and data record keeping of the bridge inspection programs.

D. Operation

In New Mexico an engineering technician inputs the data via a remote terminal with the results available immediately. If the proposed route contains one or more inadequate bridges, an alternate route can be chosen and checked by the same system or a more detailed look can be taken of the inadequate bridges.

In some cases a requirement for adjustment of axle load or axle spacing, restrictions on speed, or reevaluation of bridge capacity is all that is required to approve a permit. In other cases the vehicle may not be allowed to use one or more structures and the construction of a detour may be necessary. In all cases the engineer can be confident that all bridges on a route have been considered and the critical bridges, along with essential data, have been referred to him.

Many requests each day for permits based on gross weight are received by the New Mexico State Highway Department. All such requests are evaluated with the computer system. The majority of the single-trip requests indicate no inadequate bridges along their proposed route. Only those permits indicating inadequate bridges are referred to engineers for further evaluation. The system has reduced engineering manpower require-

```
  ***** Output Data *****

Axle Numbers     1         2         3         4         5
Axle Spacing       10.00      4.00     25.00      4.00
Axle Loads     15.00     30.00     30.00     30.00     30.00
  Overload route program developed by White-Minor, NMSU
  Sponsored by the New Mexico State Highway Dept., 73-74
  Latest Update -  July 1980

The weight formula gross limit is    74875. lbs or 80000. lbs (Formula B)
               The gross load is   135000. lbs


Please note that the following results are compared against inventory load

Br No.   90  Rt Dir: 2way
       US 666 Milepost is  58.12 DOD is  335
       Timber stringer
       58.6 M.N.Jct US66 & US666
       Span lengths       25.
       Eq Hs = 33.4   Rated HS = 11.0   Ratio =3.0

Br No. 1780  Rt Dir: 2way
       NM  44 milepost is   0.23 DOD is  425
       St  Ms  Stringer
       0.4M E of US85, Bernalillo
       Span lengths       28.       42.    28.

       Pos HS = 31.9   Rated HS = 15.0   Ratio =2.1
                         Neg HS = 34.1    Ratio Neg =2.3

Br No. 1792  Rt Dir: SBL
       US 666 Milepost is  92.93 DOD is  335
       St  SS  Truss-thru
       San Juan River in Shiproc
       Truss length = 165.00  Span lengths =  15.00
       Eq Deck HS = 34.2   Eq Flbm HS = 30.5   Eq Tru  SS HS = 34.2
       Rated HS = 15.0
       Ratio-1 =2.3  Ratio-2 =2.0  Ratio-3 =2.3

Br No. 3071  Rt Dir: 2way
       US 666 Milepost is   7.39 DOD is  335
       7.8 M N Jct US66 & US666
       Conc MS culvert
       Span lengths        10.
       Concrete box culvert     Rated capacity =15.0

Br No. 8009  Rt Dir: NSBL
       US 666 Milepost is  57.96 DOD is  335
       Pre Con Stringer
       58.2 M.N of Jct 140&US666
       Span lengths       66.
       Eq HS = 27.6   Rated HS = 21.0   Ratio =1.3

XXXXX     87  Bridges have been checked on this route  XXXXX
```

Fig. 12.2 Sample Output—Overload Route Program

ments for checking structures for an overweight vehicle permit, and has also reduced the potential for error without sacrificing engineering evaluation.

The capability for checking multiple-trip requests has created some interesting situations. For instance, a request to exceed legal load limits on a continuing basis utilizing triple trailer trucks during a three-year test for fuel economy was turned down based on the equivalent HS loading exceeding the inventory capacity of several bridges along the proposed route [4]. Similar requests involving ore-hauling trucks have been turned down [5]. The capability of checking against inventory ratings is significant in such cases.

A similar computer program has simplified the posting of inadequate bridges. Specified legal loads were passed over all bridges within the state via the computer and compared with the operating capacity. Those bridges in which the equivalent legal load exceeded the operating rating were posted at operating capacity. The system also provides a quick method for listing posted bridges.

III. Bridge Capacity Analysis

Numerous computer-aided analysis systems are available as a tool in determining the safe load-carrying capacity of bridges. These systems are valuable in determining the initial inventory and operating ratings of bridges. The systems can also be used for a detailed analysis of the bridges identified by the ØVLØAD system.

A. Bridge Rating and Analysis (BRASS)

One of the first systems developed with specific emphasis on bridges was the BRASS System [6]. The system was developed by the Wyoming Highway Department and sponsored by the Federal Highway Administration. One of the primary goals of the system was to provide a computerized method of determining the inventory rating and operating rating described in the Manual for Maintenance Inspection of Bridges [1].

The input required is the structural data from the "as construc d" plans and/or design file, the structural loading and the condition rating of the structural members. This information must include span length, cross-section dimensions, material properties, type of material, and type of structure, e.g., rigid frame, slant leg, or continuous beam

The system also includes bridge design, deck design and review, girder section design and review, and structural analysis. The types of structures that may be considered using this system include all types of layouts, continuous beam bridges, rigid frame bridges, slant leg structures, and rigid-frame box culverts. Materials may be reinforced concrete, structural steel, or timber.

The methods of analysis for beam type structures is column analogy. The method of analysis for cell structures or slant leg frames is slope deflection. Virtual work is used to calculate deflections. A Gauss-Jordan method is used to invert the resulting matrices for solution of the governing equations.

B. A Computer System for Analyzing and Rating Bridges (BARS)

The Control Data Corporation developed the BARS system to provide complete analytical capabilities for bridges. The goal of the system is to perform inventory rating, operating rating, posting ratings, special permit analysis, and analysis for bridge design. Five types of structures may be analyzed: decking, stringers, floor beams, girders, and trusses. Construction material that may be used in the analysis includes structural steel, reinforced concrete, prestress concrete, and composite girder-deck systems.

The input is typical of structural analysis systems and includes the geometry of the structure, member properties, member materials, and loadings. The analysis is based on the working-stress method described in the AASHTO specifications.

C. Bridge Analysis and Design (BRANDE)

The BRANDE system was developed primarily to analyze the grid system of bridge superstructures [7], but is also capable of analyzing rigid frames associated with either the superstructure or substructure. One of the goals realized in this system is an analytical tool that provides a fast, accurate, and detailed analysis of bridge superstructures with a minimum of input data required. The BRANDE utilizes an input format that is as simple and forgiving as the ICES-STRUDL systems and yet requires a greatly reduced number of input statements for the analysis of most bridge superstructure.

The BRANDE system is designed primarily around an elastic analysis but a plastic analysis is also available for behavioral studies of steel bridges. Three basic geometric systems are incorporated; the right and skew grids for bridge superstructures or a general configuration for grids or frames of either superstructures or substructures. Other convenient features include: potential for inputting variable member properties, various support conditions, using 3 to 6 degrees of freedom per joint, and an internal units conversion. Units may be entered in the U. S. Customary System or the International System (SI). Numerous loading possibilities are available on the BRANDE system. Loads possibilities include concentrated loads on both joints and members, uniform loads, support settlements and a deck load distribution approximation.

The results or output from BRANDE is comprised of bending moments, torsion, shear and axial forces at the member ends in addition to joint displacement and rotations. The user has the option of specifying part or all of the output for selected members, joints, or the entire structure. This feature gives user the option of generating very detailed output for any number of load cases or of selecting only a few critical members and observing the effect of various loads on these selected members.

The BRANDE system is presently being used by the New Mexico State Highway Department. One major study on bridge diaphragms was completed using the BRANDE system [8]. The system is available for day-to-day bridge analysis also.

References

1. American Association of State Highway and Transportation Officials, Manual for Maintenance Inspection of Bridges, 1974, Washington, D. C., 1974. Also AASHTO Interim Specification Bridges 1976, Revision Interim 16, p. 35.

2. J. Minor and K. R. White, A State's Solution to the Bridge Inspection Problem, American Association of State Highway and Transportation Officials, American Highways, Oct. 1972, Washington, D. C.

3. American Association of State Highway and Transportation Officials, Specifications for Highway Bridges, 1977, Washington, D. C., 1977.

4. H. Houghton, The Interstate System: Ills Plague Superhighways, Albuquerque Jounral Magazine, Impact, 1:12, 1978.

5. B. Brown, Grants Ore Hauler Trying to Win OK for Bigger Trucks, Albuquerque Journal, August 7, 1976.

6. Wyoming Highway Department, Bridge Division, System Orientation Manual for Load Rating of Bridge Structures, 1973, Laramie, Wyo.

7. W. C. McCarthy, Elastic-Plastic Analysis of Stringer-Diaphragm Systems, Master's Thesis, New Mexico State University, 1976.

8. W. C. McCarthy, K. R. White, and J. Minor, Interior Diaphragms Omitted on the Gallup East Interchange-Interstate 40, Civil Engineering Design, Vol. I, No. 2, Marcel Dekker, N. Y.

9. K. R. White and J. Minor, Evaluation of Bridge Overloads, Transportation Div., American Society of Civil Engineers, January 1979.

Chapter 13

WATERWAYS AND TERRAIN

I. Introduction

In the 15-year period, 1955-1969, an average of 22.6 million dollars per year of federal emergency relief funds was spent in financing repairs and the reconstruction of roads and bridges on Federal-aid Highway Systems and on roads and trails under federal supervision. State and local authorities also spent a considerable amount. A considerable portion of these expenditures came as a direct result of water damage on bridges or waterways.

II. Problem Areas

A. Scour

A serious problem which is frequently encountered around piers and abutments is scour. Scour is the erosive action of running water in streams in excavating and carrying away material from the bed and banks; it can occur in both earth and solid rocks.

Scour is often the cause of bridge failures that can force road closures, result in loss of life, cause excessive travel delays, and involve costly economic losses. The costs for repair and in some cases replacement of the bridges have proven to be quite high. The bridge shown in Fig. 13.1 was damaged by the scour of flood waters.

Scour as a natural phenomenon takes place primarily in alluvial streams, although scour may occur in all streams. Scour may also occur through the action of waves and tides in coastal areas, lakes, and reservoirs. Scour as a category has four subdivisions:

1. Scour occurring in a stream itself with or without a bridge.
2. Scour occurring at a bridge site due to contraction of the stream flow.
3. Scour occurring because the course of the stream flow is altered by the bridge piers and abutments.
4. Any combination of the above three situations.

Fig. 13.1 Scour Damage to Bridge Caused by Flood Waters

Stream behavior due to scour may result in displacement of the channel by migration of a meander, shift of a thalwet (deepest part of the channel), or a chute cut off (straightening of a bend). Scour may also result in the deposition of material to another area. Furthermore, the amount of scour is dependent upon the shape, size, and orientation of the pier, and the flow pattern of the stream.

B. Degradation and Aggradation

Degradation is a form of scour that occurs naturally or artificially. This scour is the result of augmentation of the stream flow, or the reduction of sediment supplied to a reach, as by a dam. The streambed may lower considerably due to degradation; therefore, the likelihood of this action should be assessed in the hydraulic design of any bridge. The vulnerability of existing bridges also should be checked when river-control works are planned that could result in degradation.

Degradation is a condition which should be examined very carefully by the bridge inspector. Conditions to look for and inspect are:

1. Channel drops and spillways in vicinity of the bridge. For example, an undermined check dam or spillway built to protect upstream structures from degradation is a potential source of trouble.
2. Changes in the underpinning of existing structures.
3. Degradation of the streambed near footings. For example, timber piles become exposed below the footings due to degradation of the stream flow.
4. Bed elevation differences should be noted over a period of time. For example, culverts are sometimes undermined by the gullying or erosive action of water. Investigate whether any bed degradation is a result of natural or artificial causes.
5. Channel shifts that have been straightened downstream can lower the streambed and result in degradation.

The opposite type of action, aggradation, in which there is deposition in the river channel, should also be assessed. The obvious problem is the raising of the water surface because of a rise in the streambed. This action would not seem to be a scour problem; however, the rise in water surface could cause the build-up of drift on the lower members of the bridge, thus restricting the water area under the bridge, and increasing the capacity of the flow locally to transport sediment.

Deposition in the river channel can also result in a greater percentage of the flood flow encroaching on the floodplain, which could result in added local scour at a constricting bridge.

C. Debris

The size of debris is always an important consideration in evaluating the adequacy of bridge clearance. Debris can sometimes be generally classified as small, medium, or large by noting the size of the vegetation washed-up along the banks and trapped against the piers.

Undercut banks with large trees near the edge can create a potentially hazardous situation for a downstream bridge.

D. Ice Damage

A bridge inspector should check for past damage to trees and bridges to evaluate the ice-carrying capacity of the stream. Also look for scarring of the tree trunks and trusses or beams bent downstream.

Fig. 13.2 Damage to Bridge by Flood Waters

E. Dredging of a Channel

Dredging upstream or downstream from the bridge site can cause bridge failures during and after such an operation. Existing bridge structures should be inspected before dredging to ensure that any increased scour does not place the bridge in jeopardy. Evidence of past dredging activity might be obtained by comparing the present bed elevation with that indicated in the original drawings or previous inspections of the bridge. Any evidence of diking, unnatural straightness, or artificial fill should be noted.

Fig. 13.3 Undermining of Abutments or Piers, Common Problem for Bridges

F. Floods and Flood Damage

During floods the discharge and velocity of stream flow is much greater than under normal conditions. For example, in May of 1945 a flood occurred on the Columbia River in the state of Washington. The discharge quantity was 13 times greater than normal and the velocity of flow was 6 times greater than normal. Floods have the capacity to transport large sediment loads which can result in increased erosion.

Fig. 13.2 provides an example of damage caused by flood waters. This bridge collapsed after the flood waters undermined the pier. Fig. 13.3 shows the damage to a small bridge due to the undermining of flood waters.

Fig. 13.4 Spring Thaw Occasionally Causes Damages Similar to That Shown Here

The bridge in Fig. 13.4 was washed out by flood waters produced by snow melting rapidly in the spring.

Records of major floods are often available from local sources and can be useful to the bridge inspector. The bridge inspector should:

1. Check past flood rises
2. Estimate flood discharges
3. Evaluate adequacy of the waterway under a bridge to handle flood discharges
4. Survey bridges after severe storms in spring and summer or after ice melting in the spring.

Some possible ways of checking or estimating past floods are from observing and recording ice scars or high water marks on tree trunks or telephone poles, trusses bent downstream from ice flow, debris, severe flooding, or debris strewn along the flood plain.

Bridges with clearances higher than their approach roads may indicate a history of flood rise was included in the design.

III. Measuring, Recording, and Detecting Scour

It is generally difficult to predict how water flow will affect the streambed in the future. However, there are many important aspects of a bridge waterway which should receive the bridge inspector's attention.

A. Useful Information

Information useful in making a bridge waterway inspection includes:

1. Location of bridge
2. Description of bridge
3. Estimated age of bridge
4. Overflow history
5. Scour damage
6. Scour holes
7. Streambed composition

To facilitate identification of a particular bridge, a brief description of its location should be the first item to note.

The description of the bridge should include type of bridge (truss, cantilever, etc.); number of spans; number, type, shape, and skew of piers; depth of footings, and type and depth of piles.

An estimated age may be obtained from the owning agency or local scources. The age is helpful in evaluating the waterway.

Overflow information is important in that water flow over approach roads during floods acts as a safety valve. A good inspector will compare how much of the flow discharge can be handled by the bridge waterway and how much can be handled by the approaches. One should note if the approach road is lower than the bridge deck so that overflow is possible. Also check for visible indications of past overflow if flood records are not available.

If scour damage is visible, any abutment movement is most likely due to scour rather than foundation problems. Check for any old structures that were washed out by previous floods. Repair work often indicates past scour problems. Examples of past repairs provide valuable clues to scour damage to various elements, such as

1. Underpinning of substructures
2. Relatively new steel sheet piling
3. Piers or abutments of different ages

4. Piers or abutments built up to correct for tilt

5. Concrete floor under bridge on the streambed

6. Stream flow; rapids caused by riprap deposits

Scour holes downstream from bridge give clues as to adequacy of the waterway under the bridge. Large downstream scour holes usually indicate that the bridge waterway is too small. Large scour holes may be indicative of high flood velocities. Absence of scour holes usually indicates that the waterway is not seriously constricted by the bridge. It must be remembered that scour holes may fill in as a flood passes. Downstream scour is accelerated if hard rock makes up the upstream beds or if the banks consist of highly erodible materials. Downstream scour is usually a result of high velocity flood flow, and the downstream scour may cover a wide area. Downstream scour may also be caused by a concrete or riprap floor placed on the streambed under the bridge. If substrata around foundation is made up of fine sand overlying a layer of hard compacted soil impervious to water, scour will be likely to occur because the fine sand overlying the hard compacted soil will scour away. In recording the presence of scour holes, the shape and position of the scour hole relative to the angle of attack and the point of maximum scour should be noted.

Streambed composition should be noted and recorded for size of material deposited on the streambed indicative of velocity of past floods. Also look for large rocks or riprap beyond scour holes. Check streambed for overlays of hard material (limestone, rock, etc.) on softer material (silt, sand, mud). Bridges founded on piles may also indicate highly erodible soil is present.

B. Frequency of Inspection

Frequency of inspection usually varies depending upon organizational policy and the susceptibility of certain bridges to scour. Determining the extent of scour at bridge foundations is generally limited to observations during periodic inspections. The Highway Research Board recommends that inspections for scour be made annually on bridge sites where known problems exist and biannually at more stable sites.

C. Methods of Inspection

The most commonly used and least expensive method of inspection for scour is the taking of soundings with a weighted line. Sonar or echo-sounding devices are sometimes used because of their great mobility. Underwater investigations can be conducted to examine firsthand the extent of scour. In cases where footings have been undermined, inspection should be performed by experienced divers. Resistance-measuring devices imbedded in the pier are also sometimes used to measure local scour.

D. Observations and Reports

The conditions of piers and abutments, when scour preventive or corrective measures are initiated, should be documented in your bridge inspection report. Photographs should be included. The extent of preparation required for the corrective treatment, and all reconstruction, or repair needed should be documented. Where continued monitoring of the substructure unit is desired, complete instructions as to frequency, type of measurements, and purpose should be explained to the responsible work unit.

IV. Terrain Problems

A. Process of Erosion

Streambed erosion is a problem in nearly every part of our country. Erosion ruins much good land by reducing its productivity or recreational value and leaves ugly scars on the landscape. Erosion can cause clogging of stream channels by depositing large quantities of silt and other fine materials taken from fields and stream banks. This action can lead to an increasing possibility of floods and impeded navigation. Bridge substructures such as abutments, piers, and retaining walls are frequently damaged by excessive erosion.

Streams are the most destructive erosive force and are influenced by a number of items:

1. Soil type
2. Size and character of floods
3. Presence or absence of trees or shrubs
4. Velocity of stream
5. Stability of river bed
6. Climatic conditions

The amount of material eroded by streams varies greatly depending on the composition of the terrain over which it flows.

B. Construction Activity and Embankment Fills

Construction activity can increase the scour potential at both new and adjacent existing structures. Channel changes, removal of stream bed materials for embankment or aggregate, large coffer dams, and temporary ramps into the stream can produce unanticipated scour. The embankment fills of the highway crossing often will create a severe contraction of the river in flood. The floodplain flow must then move laterally to the bridge opening. Where this lateral movement takes place is very important. If the flow returns to the channel largely in a reach of some length upstream from the bridge, there will be general scour over the entire waterway opening. If, however, the flow returns along the embankment there will be severe scour at the abutment and possibly out to the first or second pier, with general scour taking place downstream from the bridge.

Erosion can cause the clogging of stream channels by depositing large quantities of silt and other fine materials taken from fields and stream banks. This action can lead to an increasing possibility of floods and impeded navigation.

C. Protection Against Erosion

The principal types of protective measures may be classified by the materials of which they are constructed, the general shape of the device, or according to their function or application. These classifications include:

1. Armor
2. Retard
3. Jetty
4. Groin
5. Bulkhead
6. Baffle
7. Vegetation

The armor type of protection involves the artificial surfacing of bed, backs, shores, or embankments to resist erosion or scour. Armor is the most common type of protective device used, because (1) protection is directly in contact with the embankment requiring protection, (2) the facility may often be built within the standard right of way, and (3) there is access for inspection and maintenance purposes from the highway shoulder. Rigid types of embankment protection such as sacked concrete, asphalt paving, or mortar should be checked for cracking or misalignment by visual observation. Sight along the surface for any abnormal horizontal and vertical displacement. This cracking or misalignment sometimes results from subsidence, undermining, outward displacement by hydrostatic pressure, slide action, or erosion of the supporting embankment at the ends. If embankments are paved, weepholes should be provided to prevent damage by hydrostatic pressure or natural seepage behind the pavement. Weepholes should be reopened if the flow is impaired or new holes drilled if seepage is evident. Any cracks in the pavement surfaces should be filled with a mortar or a sealant to prevent the loss of embankment material. Coarse brush and trees should not be allowed to grow in joints or close to the edges of the structure.

Asphalt slope paving, which may be laid on slopes not previously sterilized to prevent plant growth, may be damaged or destroyed by such growth subsequent to paving. Growth through asphalt should be eradicated by chemical means, the protruding growth removed, and the surface patched.

Flexible types of embankment protection such as wire mattresses or riprap are less susceptible to damage by distortion and are intended to continue to function after moderate displacement. Wire-mesh netting should be checked for broken tie wires, missing anchor pins or cables, weak or damaged areas, or gaps between the netting and bridge structures. Rock-and-wire mattresses are sometimes placed at the base of some other type of slope protection, generally of the rigid type, when resistance of the foundation to erosion is questionable and a safeguard against undermining is warranted. If severe erosion is taking place in spite of this riprap protection, then, the protective layer should either be thickened or replaced by some other more stabilizing material. Erosion may also be due to a pan filter blanket. Examine especially the areas at the toe of the embankment to see if the embankment protection is resisting scour from water flow or wave action. It may be desirable to place excess rock in the case of riprap protection at this critical point to inhibit scour.

Retards are another type of embankment protection devices, which are designed to induce silting or trap floating debris. Retard types are permeable structures customarily constructed parallel to the toe of the slope. Their primary purpose is to offer protection to the toe by reduction of flow velocity. The resulting deposition reverses the trend of erosion and replaces material which has been lost. This causes a shifting of the flow away from the bank. The following are common types of retards:

1. Steel tetrahedrons
2. Concrete tetrahedrons
3. Timber jacks
4. Steel jacks
5. Single or double line of fence
6. Line of timber pilings or pile bents

These types of retards should be inspected to ensure that they are intact and fulfilling their function of protecting the stream bank. If rails or piling are used then they should be examined for deterioration, broken or missing sections, bracing, wales, and wire mesh. Any large items of debris that endanger the structural integrity of the retard should be removed.

A jetty is a kind of artificial wall built out into the water from the bank to restrain currents and protect the ends of piers and abutments from severe scour. Some types of jetties are impermeable (such as some log-types) and serve to deflect the stream flow. Permeable jetties also deflect the stream water but they allow some water to flow through them. Common materials used are timber-braced piling or old automobile bodies. One should also inspect jetties to make sure they are solidly anchored by cables or some other means and that they have not been laterally displaced by the current.

A groin is a bank of shore-protection structure in the form of a barrier oblique to the primary motion of water. Structures composed of steel, concrete, and timber sheet piling and cells or cribs of steel or concrete, are generally of substantial design requiring no maintenance short of reconstruction. Fill material lost from cell or crib types should be replaced.

The bulkhead or retaining wall is a steep or vertical structure supporting natural or artificial embankment. The bulkhead types are usually rather expensive, solid types, which may be economically justified in special cases where valuable stream property or improvements are involved, and the foundation is not satisfactory for cheaper types. They may be used for toe protection in combination with revetment types of protection. Included in this classification are the following:

1. Timber on timber piling
2. Concrete wall
3. Masonry wall
4. Timber sheet piling
5. Steel sheet piling
6. Concrete sheet piling
7. Concrete crib
8. Timber or log crib

Partial or complete failure of such structures may be attributed to loss of foundation support, erosion of natural banks adjoining the structure, or loss of fill material from the open crib type.

A baffle is a pier, vane, sill, fence, wall, or mound built on the bed of a stream to parry, deflect, check, or disturb the flow or to float on the surface to dampen the wave action. The fence types are the most frequently selected of the devices in this classification as they are adapted to many hazards or needs and the structural requirements can be readily provided.

The cheapest form of bank protection is an effective cover of vegetation. One of the most effective types of vegetation cover is a stand of willow trees.

Chapter 14

MAINTENANCE AND REHABILITATION

Volumes of information on maintenance and rehabilitation of bridges has been published in recent years. For this reason, this chapter is not intended to provide detailed specific information on these topics. This chapter will provide a summary of maintenance and rehabilitation activity and provide a reference source for specific areas of interest to the bridge inspector.

I. Bridge Maintenance

An effective maintenance operation for bridges must be closely associated with inspection of the bridges. Along with the inspection programs in each state; many states are advocating a concentrated effort to involve each local patrol crew in inspection and minor structure maintenance. This involvement will not only save future structure-maintenance dollars but will reduce the time and money spent on bridge inspection 2 years hence. In addition, it will involve more men in the inspection process and will increase their knowledge of structures and structure maintenance.

The 1971 issue of Guidelines for Maintenance Operations [1] published by the New Mexico State Highway Department provides an excellent section on normal routine maintenance of structures by the local maintenance patrol. The structure section includes a check list for inspecting structures, bank protection, roadway approaches, and expansion devices, and discussed numerous other items. The patrol foremen already have access to the AASHTO Manual for Bridge Maintenance, 1976 [2] and the FHWA Bridge Inspector's Training Manual [3]. Each manual has photographs and other illustrations showing deficiencies in bridges and sections addressed to what to look for during inspection and what to do about it.

It is important that deficiencies be corrected rapidly. The best way to get this done is to use local crews when possible. Most common problems encountered in bridge maintenance involve routine work or just plain housekeeping. The greatest amounts of time spent by most maintenance crews on bridges involve removal of dirt and debris, cleaning and painting, snow and ice removal, and patching or overlaying the deck.

Corrosion of steel members or steel reinforcement is one of the major causes of bridge deterioration. The corrosion process requires the presence of water to remain active. Therefore, the removal of debris from bridge deck, around bearing areas, at expansion joints, in drainage devices and on bridge pier caps is very important for retarding corrosion by reducing the capacity for water storage. Cleaning and painting of bridge members is also very important. A regular program for cleaning and spot painting of localized areas of rapid paint failure, such as beam ends under open or leaky joints, lower flanges of beams under floor drains, and similar locations will prevent corrosion and prolong the life of the entire paint system.

The Transportation Research Board (TRB) Special Report 185 [4], Snow Removal and Ice Control Research, provides an extensive study of this particular maintenance problem on bridges. Such preventive measures against corrosion are routine maintenance procedures in the cold regions of the world and the associated problems of concrete-deck deterioration due to chemicals used for snow and ice removal confronts almost every highway agency in the United States.

Bridge deck deterioration caused by chemical action on the concrete, corrosion of the steel, or simply the abrasive action of traffic makes deck overlays an important maintenance consideration. These overlays may serve to improve the riding surface or protect the concrete deck. In recent years, waterproofing systems to prevent water and chlorides from reaching the reinforcing steel have become common [5]. Such systems include latex-modified concrete wearing surfaces, low slump surfaces, and preformed rubberlike sheet membranes with an asphaltic concrete wearing surface.

Repair of traffic or collision damage is another important area of bridge maintenance. Such repair is often very difficult and may require beam or girder replacement in extreme cases. Truss bridges are particularly susceptible to traffic damage, since most are quite narrow and all the main load-carrying members are often above the roadway surface. Vehicles striking the portal or wind-bracing often pull main truss members out of alignment. If a compression member is involved, this bending may well greatly reduce its buckling resistance. Even minor nicks and gouges in steel members may cause stress concentration locally and often propagate cracks.

Bearings are used to transmit and distribute superstructure loads to the substructure and permit the superstructure to undergo necessary movements without developing harmful overstresses. Routine maintenance is often required to insure that bearing devices work properly. A build-up of debris and loss of paint are common bridge-bearing problems. A routine program to clean off dirt and debris and check joints to insure that joint seals are working properly can improve the effectiveness of bearings. Any sheared or corroded anchor bolts, retainer plates, or similar devices should be replaced by bridge maintenance crews.

II. Bridge Rehabilitation

The dividing line between bridge repair or maintenance and bridge rehabilitation is rather hazy. In the 1978 Bridge Condition Rating Guide [6] descriptions, the difference is implied to depend upon the extent of repairs required to bring the bridge up to an adequate condition. Rehabilitation is usually intended to extend the service life of an existing bridge until such time as money, time, and manpower are available to replace the structure. Several research projects have been sponsored in recent years on the rehabilitation of bridges. These projects include research on increasing the load carrying capacity of bridges [7] and rehabilitation of "off-system" bridges [8].

Rehabilitation may take the form of any one of several restoration procedures. The work can include deck replacement and minor repair or it can involve procedures for correcting settlement problems, strengthening or replacing critical members, replacing bearings, widening or correcting alignments, or improving drainage [9].

A. Deck Replacement

The most common rehabilitation of a bridge is replacement of the deck or of the deterioration portions of the deck. The deterioration is commonly caused by chloride penetrating concrete and corrosion of the reinforcement as a result primarily of de-icing chemicals placed on the bridge deck. The type and extent of deck restoration depends greatly upon the chloride content and percentage of deck area contaminated [10]. Often bridge decks with less than 1 lb of chloride per cubic yard at the rebar level are protected by overlaying with a waterproofing membrane or low slump concrete. For bridge decks with greater than 2 lb of chloride at the rebar level (the critical salt concentration), most highway agencies remove the contaminated concrete to below the reinforcing bars, sand blast the rebars, coat the rebars with an epoxy protection material, and pour new concrete.

For bridge decks with extensive chloride contamination, cathodic protection and epoxy grouting are sometimes used. Cathodic protection systems have been used for several years now by a few state agencies on decks with advanced rebar corrosion over significant areas of the bridge deck. Complete replacement of bridge decks is normally recommended if more than 40% of the surface area of the deck is contaminated. In some cases, temporary patching and pothole repair is used on badly contaminated decks until the corrosion of the rebars or concrete deterioration renders the structure unsafe for legal loads to cross the structure. Normally the deck replacement incorporates coated rebars in the upper layer and a waterproofing membrane or low slump concrete-surfacing to protect the deck from early chloride contamination.

B. Girder Replacement or Strengthening

Several methods are available for repair of concrete bridge girders depending on the type and extent of repair needed. Some cosmetic patching is used to protect exposed rebars but as a general rule gluing concrete pieces together has not been successful [11]. Epoxy injection can be successfully used to protect reinforcement, particularly prestressed, tendons from moisture. Such application should only be used, however, after the structure has been reanalyzed to take into account any other deficiencies caused by deterioration. This type of rehabilitation may be considered only maintenance since the structure is not strengthened and the process simply slows down the deterioration and temporarily prolongs the service life.

Other concrete-member rehabilitation includes external steel reinforcement attached to the member by utilizing bolts extending through the member [9]. This external reinforcement can also be post-tensioned if desirable. Supplemental steel or precast concrete members may also be used to increase the overall capacity of a bridge particularly if only the floor or deck system in inadequate. Replacement of damaged or critically deteriorated members has been a standard practice by many road agencies. This procedure usually requires careful planning to provide satisfactory results.

Steel members can usually be strengthened by adding cover plates or web plates depending on the critical stress mode. Generally these plates are bolted to prevent stress raisers susceptible to fatigue failure. Steel members usually can be replaced if damage or deterioration render them ineffective. Replacement of members in a steel truss is common practice. Steel girder bridges may be strengthened also by composite action. This improvement is accomplished by providing a shear connection between a concrete deck and the steel beam. During the replacement of deteriorated decks, studs may be welded to the beams to provide the necessary shear transfer for composite action. Other methods of accomplishing such shear connection are drilling holes through the deck for attaching studs and epoxy injection between beam and deck [9]. Splicing together simple steel beams for continuous action may also be used to strengthen existing steel bridges.

Timber bridges are probably the easiest to strengthen. One method is to simply add supplemental members of timber or steel to reduce the load for each member. Temporary bents are used by some road agencies to reduce the span on existing timber bridges. This type of action is usually taken only to extend the life of the structure for a short period until a replacement structure can be funded and designed.

C. Dead Load Reduction

Another method of increasing the load-carrying capacity, which is often easily accomplished, is the reduction of the dead load. In many older bridges the asphalt overlays have built up until the dead load from this material is significant. In some cases, the increase capacity can be accomplished by simply removing excess overlay material. In other situations, the entire deck may be removed and replaced by a lighter weight decking material.

Three materials used for new decks on older bridges are (1) the open grid steel flooring, (2) corrugated metal plates with asphalt, and (3) laminated timber decking. The open grid has the advantage of letting rain and snow pass through and eliminating the need of a deck drainage system, but the steel may become slippery when wet or ice-covered. The corrugated plate system is placed over existing stringers and some supplemental lightweight support beams. Here, the drainage system must be adequate to remove water or the corrugated metal plate will corrode. The laminated timber deck is fairly new. Prefabricated panels are normally clamped or bolted to existing girders. A properly laminated panel provides advantages in that it is resistant to chemicals used in snow and ice control.

D. Geometry

Rehabilitation of a bridge may include improvement of the geometry in the form of vertical clearances, widening of the structure, and horizontal or vertical alignment.

One of the most common forms of bridge damage is vehicular collision damage resulting from vertical clearance restrictions. This damage is particularly common when one or two bridges on a route have significantly less vertical clearance than the other structures. Thru trusses often fit into this category. Renovation may be accomplished by reducing the depth of portals or sway bracing or by lowering the floor system to increase the vertical clearance. A thinner deck system may also provide some additional clearance. Lowering the roadway at grade separations has been used by some road agencies to improve the vertical clearance.

Numerous methods of roadway widening have been used by the various highway agencies in the United States. The Hackensack River Bridge was widened by extending the structure symmetrically about the center line. The John Harris Bridge was widened all to one side [12]. The life of many bridges can be extended appreciably by widening of the structure to meet the minimum standards for today's traffic. The process is fairly routine on most bridges and involves removal of sidewalks and curbing, extending piers and abutments, and adding new stringers and a new deck.

E. Mechanical Deficiencies

The bearings, expansions, hangers, and similar devices associated with structural contraction or expansion frequently need rehabilitation. These devices often cease functioning properly as a result of corrosion or debris build-up. Usually, rehabilitation involves cleaning these devices and adjusting to the proper position. Minnesota has initiated a program of removing frozen bearings, sand blasting or cleaning the devices, and providing grease inserts such that the bearings may be serviced on a regular basis [13].

Expansion joints frequently contribute to major corrosion problems on bridges by allowing water and chloride solutions to leak through to girders or pier caps. Replacement of joint seals can often improve this situation. Drainage collection systems may be installed, or repaired if already existing, to carry drainage from open joints away from bearings or hangers. Scupper may also be used to control the drainage runoff from the deck.

F. Safety and Serviceability

A bridge may be rehabilitated in several safety areas which include replacement of inadequate bridge railing and alteration of bridge railing or guard railing ends. Alteration of parapets and protection using attenuators at hazardous ends of trusses, piers or gore areas may accomplish safety rehabilitation of a bridge. Adjusting roadway alignment can have substantial impact on the safety record of a bridge.

The serviceability of a roadway can be improved in several ways. Repair of approach-slab settlement at the end of bridges can improve the riding quality considerably for many structures. Deck repairs of pot holes and slippery areas can also improve the safety and riding comfort of a bridge.

References

1. New Mexico State Highway Department, Guidelines for Maintenance Operations, Maintenance Division, Sante Fe, New Mexico, 1971.

2. American Association of State Highway and Transportation Officials, Manual for Bridge Maintenance, Washington, D. C., 1976.

3. Federal Highway Administration, Bridge Inspector's Training Manual, Washington, D. C., 1971.

4. National Academy of Science, Snow Removal and Ice Control Research, Special Report 185, Transportation Research Board, Washington, D. C., 1979.

5. P. J. Deithelm, R. G. Tracey, and R. C. Ingberg, Bridge Deck Deterioration and Restoration, Minnesota Department of Highways, St. Paul, Minnesota, 1972.

6. Federal Highway Administration, Recording and Coding Guide for Structural Inventory and Appraisal of the Nation's Bridges, Department of Transportation, Washington, D. C., 1979.

7. R. H. Berger, Extending the Service Life of Existing Bridges by Increasing their Load Carrying Capacity, FHWA Research Report, 1978.

8. Virginia Highway and Transportation Research Council, Bridges on Secondary Highways and Local Roads—Rehabilitation and Replacement, National Cooperative Highway Research Program, p. 20-25.

9. R. H. Berger, and S. Gordon, Extending the Service Life of Existing Bridges, Transportation Research Record 644, Transportation Research Board, St. Louis, 1978.

10. G. Dallaire, Halting bridge deterioration on existing bridges, Civil Engineering Magazine, New York, October, 1973.

11. H. P. Koretzky, What Has Been Learned from the First Prestressed Concrete Bridges—Repair of Such Bridges, Transportation Research Record 664, Transportation Research Board, St. Louis, 1978.

12. M. H. Soto, Some Considerations in Widening and Rehabilitation of Bridges, Transportation Research Record 664, Transportation Research Board, St. Louis, 1978.

13. D. Flemming, Technical discussion on bridge rehabilitation. Presented at the Annual New Mexico Bridge Inspection School, Las Cruces, New Mexico, May 1977.

INDEX

Abutments, 33, 34, 127, 131, 149
Aggradation, 234
Air buffers, 163
Alignment, 96
Alluvial soil, 124, 125
American Association of State Highway and Transportation Officials, 3
American Society of Civil Engineers, 3
Analysis, 229
Arch deck, 8
Arch thru, 8
Arches, 16, 33
Arterial, 7
Axial forces, 49

Baffles, 244
Bailey bridges, 29
Baltimore truss, 14
Barges, 189
BARS, 230
Bascule, 26, 157, 165, 170
Beams, 73, 90, 102, 103, 106
Bearing capacity, 129
Bearing plates, 105
Bearings, 105, 106, 108
Bending forces, 49, 50
Bents, 34, 120, 138
Bolts, 19, 99, 100, 103
Box beams, 8
Box culverts, 27
Bracing, 33, 88, 100, 105
BRANDE, 230
Bridge maintenance, 1
Bridge pier, 127
Bridge protective systems, 134
Bridge rating, 194, 220
Bridges, 113, 124
Buckling, 49
Buffers, 169
Built-up beams, 99
Bulkheads, 244

Cable stayed, 29
Cables, 102, 164, 167
Caissons, 134
Calipers, 188
Camel, 14
Caps, 34, 139
Cast-in-place, 22
Centering blocks, 162
Circular arch, 16
Clay, 125
Clusters, 134
Coarse-grained soil, 126
Cohesive soils, 127, 130
Collector, 7
Collision, 48
Columns, 17
Composite, 20
Composite beams, 99
Compression, 13
Computers, 223
Concrete, 44, 142
Concrete beam, 22, 96, 110
Concrete boxes, 26
Concrete bridges, 7
Concrete continuous bridge, 7
Concrete decks, 72, 83

Concrete girder, 96, 205
Concrete slabs, 199
Connections, 75, 79, 99
Continuous span bridge, 227
Control house, 171
Corrosion, 113
Counterweights, 162, 167
Cracking, 105
Cross section, 74, 79
Crushing, 92
Culverts, 8, 144, 145, 227
Curbs, 114, 118

Damage, 96, 138, 166
Dashpot hydraulic, 136
Dead loads, 47, 102
Debris, 146
Deck, 65, 71, 82
Deck loads, 87
Deck system, 29
Deck truss, 14
Defects, 166
Deflections, 102, 113
Degradation, 234
Deterioration, 87, 96, 120, 138, 166
Diagonals, 12, 33
Diaphragms, 102
Direction of route, 61
Distribution, 212
Divers, 143, 189
Dolphins, 133, 138
Drainage, 168
Dredging, 236
Dyes, 188

Earthquake, 48
Elastomeric bearing packs, 110, 113
Elliptical arch, 16
Energy adsorption, 135
Engines, 169
Equipment, 167, 185
Equivalent loading, 223
Erosion, 148, 242
Erratic soil mass, 126
Evaluation, 223
Excessive wear, 120
Expansion joints, 119
Expressway, 7

Failure, 35
Fasteners, 113
Fatigue, 36, 42, 167
Federal Highway Administration, 1
Fenders, 133, 137, 138
Fine-grained soil, 126
Floating fenders, 134
Floods, 237
Floor beam system, 8, 30, 72, 87, 89
Floor system, 29
Forces, 49
Forms, 66
Foundation soils, 124, 125
Fracture, 120
Frame, 8
Freeway, 7
Friction, 111

Gates, 163
Geometry, 250
Girders, 8, 19, 73, 90, 102, 106, 110, 162
Glacial soils, 124, 125
Gothic arch, 16
Gravity fenders, 136
Groin, 244

Hanger-type connection, 108
Hangers, 100, 113
Hardpan, 126
Highway structures, 2
Hinge bearing, 107
Home truss, 13
Hydraulic buffers, 163

Ice, 48, 235
Insects, 39
Inspection, 1, 2
Inspection frequency, 4
Inspection personnel, 4
Inspection requirements, 4
Inspector, 124, 141
International Bridge, Tunnel, and
Turnpike Association, 3
Interstate, 7
Iron bridge, 16

Jetty, 244
Joints, 100

Knee bracing, 104
K-truss, 14

Lateral shear, 113
Lighting, 175
Liquid limit, 127
Line loads, 47, 102, 212
Load factor, 195
Load flexible, 136
Loads, 31, 47, 73, 87
Locking devices, 163

Machinery, 168
Magnetic particles, 190, 192
Maintenance, 246
Marine borer, 39
Masonry, 7
Measurements, 5
Mechanics, 47
Menai Strait, 20
Moments, 51, 53
Motors, 169
Moveable-bascule, 8
Moveable bridges, 26, 156
Moveable-life, 8
Moveable-swing, 8

Noncomposite, 22
Nondestructive testing, 190
Notebook, 62

Organic soils, 126
Orthotrophic, 8

Pads, 110
Parabolic arch, 16
Particle size, 126
Pedestal-and-shoe bearing, 107
Photographs, 69
Pier, 33, 34, 138, 151, 168
Piles, 130, 141
Pins, 113
Plate girders, 19
Plates, 110
Pneumatic, 136
Pontoon, 29, 160
Pony truss, 14
Portal bracing, 102
Pratt truss, 19
Precast, 23
Prestressed, 23
Properties of soil, 126

Radiographics, 190, 191
Railings, 114, 117
Rating, 5, 68
Reactions, 51
Records, 4
Redundancy, 36
Rehabilitation, 246, 248
Reinforced concrete, 89
Reinforced steel, 87
Residual soil, 124, 125
Retractable fenders, 135
Rivets, 19, 99, 100, 103
Rocker bearing, 108
Rockers, 110
Rods, 102
Roller bearing, 108

Rollers, 110
Roman arch, 16
Rubber fenders, 135
Rust, 105

Safe-load capacity, 1
Safety, 181
Sagging, 89
Scaffolding, 189
Scheduling, 80
Scour, 35, 233
Seeping, 130
Settlement, 127, 128, 130
Shear forces, 49, 50, 51, 53, 92, 127
Sheet piles, 134
Sidewalks, 114, 118
Signs, 173
Simple span bridges, 227
Sketches, 62, 64, 65
Skew, 162
Skin friction, 130
Slahs, 8, 25, 103
Sliding, 130
Snooper, 189
Soil foundation, 124
Sole plate, 106
Spalling, 96, 105, 120
Spandel, 17
Spans, 8, 74
Stayed girder, 8
Stay-in-place forms, 72
Steel, 41, 89, 99
Steel beams, 99, 100, 203
Steel continuous bridge, 7
Steel decks, 83
Steel girders, 99, 100
Steel piling, 142
Steel spring fender, 137
Stiffners, 20
Strain measurement, 190, 193
Stress analysis, 216
Stress concentrations, 42
Stresses, 78
Stress points, 50
Stress strain, 41
Stringer, 8, 89, 103
Struts, 103

T-beams, 8, 22, 87
Techniques, 5
Tension, 13
Terrain, 233
Thru truss, 14
Tied arch, 17
Timber arch, 19
Timber decks, 82, 201
Timber pile system, 134, 135
Timbers, 7, 38, 88
Torsional forces, 49, 50
Transverse bracing, 102
Truck loads, 212
Truss, 13, 75, 76, 102, 106, 227
Truss deck, 8
Truss types, 31
Twisting, 89
Types of bascule bridges, 158, 165, 170
  Chicago, 158
  Roll, 158
  Scherzer, 158
  Strauss, 158
Types of bearings, 108
  Expansion, 108
  Fixed, 108

Ultrasonics, 190, 191
Unified Soil Classification, 126
Uniform soil, 126
Utilities, 178

Vegetation, 245
Vertical lift, 26, 156, 160, 170
Verticals, 13

Warning devices, 168
Warren truss, 13, 14

Water pressure, 48
Waterways, 233
Wearing surface, 86
Welds, 19, 100, 103
Well-graded sand, 125
Wind, 102
Wood piling, 142